32

# A Better Mousetrap: Improving Pest Management for Agriculture

Michael J. Dover

D1300733

WORLD RESOURCES INSTITUTE
A Center for Policy Research

Study 4
September 1985

WITHDRAWN

LORETTE WILMOT LIBRARY
NAZARETH COLLEGE

WITHDRAWN

632.9 Dov
Dover.
A better mousetrap.

The Lorette Wilmot Library
Nazareth College of Rochester

632.9
Dox

**Acknowledgments**   This paper reflects my eleven years' experience in integrated pest management and two years of study at World Resources Institute. It also reflects the ideas of hundreds of people cited in the footnotes. Many more contributed directly. Kathleen Courrier shortened and sharpened the text, and Richard Wiles at WRI helped gather material for several tables and provided valuable comments on an earlier draft. Janet Brown, Peter Thacher, and Monty Yudelman of WRI; George Bird of Michigan State University; Richard Harwood of International Agricultural Development Service; David MacKenzie of Louisiana State University; and David Pimentel of Cornell University all made cogent and useful recommendations. Gus Speth, Jessica Mathews, and Rob Wasserstrom also provided much-appreciated direction and support.

*M.D.*

**Foreword**    Scores of Americans spent this Fourth of July in emergency wards—innocent victims of pesticide residues. A few days after the poisonings, California state authorities ordered growers, wholesalers, and grocers to destroy more than a million striped watermelons like those that caused the poisonings. Aldicarb, a chemical banned for use on melons in this country, had made forbidden fruit out of a summertime favorite and ruined at least 5 percent of the 1985 watermelon crop.

Pesticide abuse is but one of the many problems associated with reliance here and abroad on a "chemicals-only" approach to pest control. Increasingly, scientists, policymakers and others are asking how much is enough to do the job, what constitutes overkill, what pesticides do over time to human tissues, and how they affect the environment. And scientists and policymakers alike are trying to figure out which of the new generation of pest-control technologies, chemicals, and approaches offer the safest protection now.

In *A Better Mousetrap: Improving Pest Management for Agriculture*, Michael Dover takes a hard look at this last question and finds some hopeful answers. Tremendous and growing demands on world agriculture have all but extinguished pastoral dreams of chemical-free farming, but they have also spurred the development and limited use of "better mousetraps": safer and more effective devices for manufacturing, packaging, storing, and applying pesticides, as well as controlling their drift. Most important, they have given impetus to efforts to improve biological methods of pest control—using, if you will, bugs to fight bugs—and to develop system-wide approaches to the never-ending war with vermin.

As Dr. Dover points out, systems management at its best is Integrated Pest Management (IPM), an arsenal of tools and techniques first developed in the 1960s. Two decades ago, evidence began accumulating that using pesticides indiscriminately entailed far greater environmental hazards, health risks, and costs than their manufacturers and users had imagined. Shelving hopes for the one-shot miracle chemicals that farmers call ''magic bullets,'' advocates of IPM began successfully combining regulation, safety training and worker certification, research, and demonstration programs with local or regional needs in mind. Today, no two IPM programs are identical, but all try to minimize risks and costs while maximizing effectiveness and profits—a goal that farmers, chemical manufacturers and environmentalists will all support if given the right information and the right signals from government.

*A Better Mousetrap* is the third policy study to grow out of two years of intensive research at World Resources Institute on pesticide issues. The first, *Getting Tough: Public Policy and the Management of Pesticide Resistance* by Michael Dover and Brian Croft, put the pesticide-resistance problem into perspective and outlined a seven-part U.S. strategy for controlling unwanted immunity to pesticides. The second, *Field Duty: U.S. Farmworkers and Pesticide Safety* by Robert F. Wasserstrom and Richard Wiles, explored the scientific and policy issues surrounding pesticide use and farmworker safety, offering a framework for systematically reducing the occupational hazards of pesticide use.

World Resources Institute sponsored these studies to provide industry, agriculture, and government with perspectives on the latest scientific information on pesticide use in the United States and abroad. All three reflect WRI's commitment to the wiser, more cost-effective, and inventive use of natural resources. We gratefully acknowledge the John D. and Catherine T. MacArthur Foundation for its support of WRI's overall policy research program and the BankAmerica Foundation for its help in meeting the expenses of printing and distributing *A Better Mousetrap*.

James Gustave Speth
*President*
World Resources Institute

**Contents**

**Figures**

**Tables**

# I. Introduction:
# The Pesticide Dilemma

In the struggle for survival, *Homo sapiens* has always had to contend with microbes, plants, and animals that threaten our health and comfort, damage our property, harm our domesticated animals, and vie with us for food and fiber. Settlement patterns, population movements, economic arrangements, food production and gathering activities, and even wars have been influenced by pests. Only in the last four decades has the relationship between people and pests shifted dramatically, raising the expectation that the battle could at last be decisively won. The revolution that has caused this change is the development, beginning with DDT during World War II, of synthetic organic chemical pesticides.

Once DDT, with its low toxicity to humans and its high toxicity to insects, was seen as a "miracle" chemical, the "perfect" pesticide. Compared to the highly toxic arsenicals, heavy metals, nicotine, and cyanide, DDT clearly was safer and more consistently effective. Long persistence in the environment and toxicity to a broad spectrum of insect species seemed like assets, given the limited effectiveness of earlier pest-control efforts. And DDT worked: it prevented millions of deaths from typhus during and immediately following the war, and it virtually eliminated malaria from large parts of the world. The introduction of DDT and the subsequent discoveries of a burgeoning chemical industry also revolutionized agriculture, helping to set the stage for substantial changes in farming practices—among them, high-yielding varieties, fertilizers, irrigation, machinery, and the

**As DDT led the way into the new age of pest control, it also became the symbol and the principal villain in the panoply of problems that emerged soon after the shift to almost unilateral reliance on chemicals for pest control.**

reduction of on-farm labor—that have profoundly affected all the world's economies and societies.

As DDT led the way into the new age of pest control, it also became the symbol and the principal villain in the panoply of problems that emerged soon after the shift to almost unilateral reliance on chemicals for pest control. Virtually simultaneously, questions began to surround the effects of these chemicals on human health, risks to the environment, and control problems springing from pest resistance and other factors.

## Health Effects, Exposure, and Safety

By design, pesticides are biologically active and, in most cases, toxic. Thus, they pose potential risks to human beings. How real and substantial those risks are depends on (1) what happens if the chemical reaches organs, tissues, or cells; (2) how much of the chemical is required to show such effects; and (3) how likely the chemical is to reach these sites in sufficient quantities to produce the effects.[1]

Forty years of experience have provided remarkably little consistent data on how synthetic organic pesticide materials affect people. Of most immediate concern is acute poisoning—sickness and death from one or a few exposures to a chemical. Estimates of the numbers of poisonings vary widely, the validity of the estimates is open to question, and the causes of the incidents are often disputed, but estimates of 10,000 deaths and 400,000 illnesses per year worldwide are probably close to the mark.[2] (See Table 1.) Even greater uncertainty surrounds the effects of chronic occupational exposure. Continuing exposure to low levels of pesticides puts certain workers at risk of cancer, tumors, reproductive disorders, birth defects, and other long-term illnesses. (See Table 2.) Those who mix, load, and apply pesticides run the greatest risk, but fieldhands and others also encounter significant levels of pesticide. Typically, these people are exhorted to handle chemicals more carefully and to use protective clothing and equipment, but improved methods of application and ways of reducing the total amount of pesticide applied can also help lower workers' long-term risks. Despite the great uncertainties of determining actual levels of chronic and acute effects, one conclusion is unavoidable: pesticide poisoning is a problem of serious proportions

**Continuing exposure to low levels of pesticides puts certain workers at risk of cancer, tumors, reproductive disorders, birth defects, and other long-term illnesses.**

## Table 1. Estimated Numbers of Poisonings from Pesticides

| Area | Date | Cases | Deaths | Source |
|------|------|-------|--------|--------|
| Worldwide | 1973 | 500,000 | 5,000 | World Health Organization (WHO) |
| Worldwide | 1982 | 750,000 | | Bull, *A Growing Problem* |
| Worldwide | 1983 | 1,500,000-2,000,000 | | U.N. Economic and Social Council |
| Cyprus | 1972 | 2.3 per 100,000 | 3.1 per million | WHO |
| Finland | 1974 | 2.8 per 100,000 | 3.4 per million | WHO |
| Rumania | 1974 | 13.0 per 100,000 | 14.4 per million | WHO |
| Syria | 1971 | 16.3 per 100,000 | 25.6 per million | WHO |
| Turkey | 1974 | 4.3 per 100,000 | 4.0 per million | WHO |
| United Kingdom | 1973 | 0.2 per 100,000 | 0.3 per million | WHO |
| Sri Lanka | 1973-8 | 14,396 per yr. average | 988 per yr. average | U.N. Economic and Social Council |

## Table 2. Chronic Effects of Pesticides as Indicated by U.S. Regulatory Actions

| Pesticide | Regulatory Action | Reason |
|-----------|-------------------|--------|
| Aldrin/Dieldrin | Canceled for all uses except termite control. | Environmental persistence and oncogenicity |
| Chlordane/Heptachlor | Canceled for all uses except termite control by certified applicators wearing protective gear. | Environmental persistence and oncogenicity |
| Chlordimeform | Voluntarily withdrawn from general use. Reregistered for use only on cotton by specially trained applicators wearing protective gear. Closed systems required. | Oncogenicity |
| Endrin | Most uses canceled; other uses voluntarily withdrawn. | Oncogenicity, teratogenicity, and reduction of nontarget species |
| EPN | Use as a larvicide canceled; protective clothing required for other uses. | Delayed neurotoxicity and acute toxicity |
| Lindane | Use in vaporizers and as a fumigant canceled; other uses restricted to certified applicators wearing protective clothing. | Oncogenicity, fetotoxicity, and acute toxicity |
| Mirex | All uses canceled except ant control on pineapples in Hawaii. | Environmental persistence and oncogenicity |
| DBCP | All uses canceled. | Reproductive effects and oncogenicity |
| Dimethoate | All dust formulations canceled; remaining uses restricted to certified applicators wearing protective gear. | Oncogenicity, mutagenicity, fetotoxicity, and reproductive effects |
| EDB | All uses canceled. | Oncogenicity, mutagenicity, and reproductive effects |

*Source:* U.S. Environmental Protection Agency.

**One conclusion is unavoidable: pesticide poisoning is a serious problem that can be diminished through better management. Substituting less toxic chemicals or nonchemical techniques for more toxic ones and preventing or minimizing pesticide exposure would both reduce poisonings.**

which can be diminished through better management. Substituting less toxic chemicals or nonchemical techniques for more toxic ones and preventing or minimizing pesticide exposure would both reduce poisonings.

Those who handle pesticides are not the only group at risk. Pesticide residues in food and drinking water can be transferred to human tissues, where the chemicals may accumulate. Cancer and other diseases due to chronic exposure may result. Especially disturbing are the long-term effects of those residues on infants and children, a subject that has received far too little attention from researchers and regulators.[3] Predicting the incidence of such diseases as a function of the level of usage involves tremendous uncertainties, debatable assumptions, and controversial judgments. Nevertheless, dietary exposure to pesticides remains a powerful social and political issue because so many people are potentially at risk and because exposure is involuntary. Indeed, even though the risk may be low compared to other environmental risks, it gives rise to the most public pressure to ban certain pesticides. Prohibiting the use of certain pesticides can be effective. Since many of the persistent organochlorine insecticides were removed from general use in the U.S., residues of these chemicals in food have dropped significantly.[4] Short of banning, however, there are other ways to reduce residues in food and water, and there is much room for additional research.

## Ecological Effects

Once released into the environment, DDT and most other chlorinated hydrocarbons can take years to break down into simpler, less toxic chemicals. And because they are more soluble in fat than in water, these pesticides tend to accumulate in living organisms. In fact, public outcry over the residues of DDT and its relatives in fish and the chemicals' effects on the reproduction of birds (due to thinning of eggshells) contributed to the decision to ban DDT for most uses in the United States and elsewhere. (DDT residues in human tissue, and the determination that the chemical is a weak carcinogen, were also major factors.)

Chemicals less persistent than the chlorinated hydrocarbons also have their deleterious effects.

Organophosphate and carbamate insecticides, for instance, are more water-soluble and thus more likely to reach aquatic and marine organisms—unintended victims of the original pesticide application. Many of these materials are also more acutely toxic, posing a greater hazard to farmworkers in particular.[5] Moreover, some persist in the environment much longer than originally thought, making longterm effects on "non-target populations" a serious possibility.[6]

## Sustainability

With the increased use of synthetic organic chemicals for pest control, three problems have emerged to threaten our ability to control pests in many parts of the world. The most significant is the accelerating development of resistance to pesticides among arthropods (insects, mites, and ticks), rodents, bacteria, fungi, and weeds. Resistance threatens farmers' ability to continue growing cabbage in Malaysia, potatoes in the Northeastern United States, and cotton in several parts of the world. It also impedes the control of insect-borne diseases in the tropics, where some pest-control programs appear to be at the end of the line because the chemical industry can no longer come up with substitutes as quickly as the pests grow resistant to the available pesticides.[7]

In both agricultural environments and natural habitats, pest populations are often kept within bounds by predators, disease, or competitors for food, water, and shelter. Sometimes, pesticide applications disrupt these natural controls, so that once-harmless species grow numerous enough to become pests. At worst, these secondary or "induced" pests cause more trouble than the primary pests against which the chemicals were aimed. Loss of natural controls can also enable pest populations to rebound quickly after a pesticide application to even greater numbers because nothing keeps them in check.[8]

**Together, resistance, secondary pest outbreaks, and resurgence often lead to the "pesticide treadmill."**

Together, resistance, secondary pest outbreaks, and resurgence often lead to the "pesticide treadmill": more and more chemical has to be applied to keep from losing everything. Just such a treadmill devastated cotton-growing in Central America. There, the number of pesticide applications per growing season increased from fewer than five to 28 in a little more than ten years, while the number of insect pest species requiring

5

control went from two to eight. Eventually, pest-control costs came to account for half of all production costs in this system.[9]

As the pesticide treadmill accelerates, costs rise from two causes. Obviously, costs go up with the number of applications. In addition, when resistance occurs or new pests emerge, new pesticides have to be purchased, often at a higher cost per application.

## The Challenge

If agricultural productivity is to be increased to meet rising food demand, and public health improved in tropical areas where insect-borne diseases remain serious problems, the stability of pest control must be a global concern. Since most productive land is already cultivated, and many wilderness areas are threatened by the expansion of agriculture, land already under the plow must be made as productive as possible. Improving pest control, both pre- and post-harvest, is one important means of raising net yields.[10] For example, up to one third of Asian rice production may be lost to insects annually,[11] and losses in fruit and vegetable crops from plant diseases in the United States could reach 20 percent.[12] Even with dramatic increases in pesticide use in recent decades, however, the percentage of crop lost to pests has apparently not declined.[13] Clearly, just pouring on more chemicals is no answer.

Pesticides present a dilemma: as their hazards become more apparent, so does the need to use them. Although designed to kill, they are often life-savers. Although increasingly costly, they bring economic benefits. And while they have opened up many possibilities for improving agriculture and public health, they have closed others, making us extremely dependent on them for our continued survival.

Can we resolve this dilemma? Can we reduce the risks of pesticide use while retaining the benefits? Must we rely solely on chemicals for effective, affordable pest control? This paper, an examination of the technologies that can be brought to bear on pest and pesticide management, shows that we can better manage pests without excessively polluting the earth and ourselves. Some of these solutions involve more carefully deciding how, when, where, and how much pesticides are applied. Some call into question the knee-jerk resort to chemical controls.

> Pesticides present a dilemma: as their hazards become more apparent, so does the need to use them. Although designed to kill, they are often life-savers.

An inherent problem in surveying new technology is the fascination with the "technical fix." Yet, the value of technological innovation cannot be denied. For instance, using better pest-management techniques could reduce pesticide use by as much as 50 to 75 percent, according to some estimates. The key lies in the context in which technologies are conceived, developed, evaluated, and implemented. The same method can produce different results, can succeed or fail, depending on how well it fits with other aspects of the management system. Is a given technology effective? That is often the principal question asked. But asking "Is it safe? Where will the technique be applied? Who will use it? How will safety be determined? What ecological systems are at risk?" broadens the context. In the end, the issues of efficacy and safety can be addressed only if a given technology is seen as a component of a system, not as an independent "free body." Pest-control methods themselves are not inherently safe or risky: the way the method is applied determines risk.

**Pest-control methods themselves are not inherently safe or risky: the way the method is applied determines risk.**

The ways in which pests and crops interact with each other and the rest of the environment are seemingly endless, especially considering local variations in weather, climate, soils, and topography. Determining the best pest-management strategy for a given site is largely a matter of figuring out how all these factors fit together there, not what worked elsewhere. What *can* be transferred from place to place is a way of thinking. How should pesticide-use decisions be made? If pesticides are needed, which one should be used and how? What alternatives are available to protect the crop? When is one option more effective, safer to people, more durable, or less harmful to wildlife than another? Questions like these, and the means for answering them, are the universal currency of pest management.

# II. Applying Pesticides: Is There a Better Way?

If chemical pesticides are likely to remain the mainstay of pest-management technologies for at least the next two decades,[14] what key issues surrounding their use need to be considered? First, the market life of pesticides can be quite long, especially when a decline in one market can be offset by opening or expanding others. Consider too that the rate of introduction of new chemical types appears to be slowing,[15] so "older" chemicals are likely to be with us for some time. Second, the newer pesticides are biologically active at much smaller doses than their older counterparts.[16] Most are also less toxic to mammals. (See Table 3.)

To the extent that less toxic compounds are either unavailable or uneconomical as substitutes, use of older, more hazardous pesticides will continue and the only way to reduce risk is to reduce exposure. In practice, this boils down to five overlapping strategies:

1. Reducing the amount of pesticide per application,

2. Reducing the number of applications,

3. Using protective clothing and other safety gear,

4. Training users in the safe handling and application of pesticides, and

5. Changing the way the pesticide is applied.

## Table 3. Relative Toxicities of Different Insecticide Classes

| Insecticide | Rat Acute oral LD50 (mg./kg.) | Insect Topical LD50 (mg./kg.) | | Toxicity Ratio rat:insect |
|---|---|---|---|---|
| Aldicarb | 0.9 | 10 | (aphids) | 0.09 |
| Disulfoton | 8.6 | 7.5 | (aphids) | 1.1 |
| Parathion | 3-6 | 0.9 | (flies) | 5 |
| DDT | 118-250 | 10 | (flies) | 18 |
| Carbaryl | 850 | 4 | (mosquitoes) | 212 |
| Dimethoate | 200-300 | 0.7 | (flies) | 357 |
| Malathion | 1400-1900 | 18 | (flies) | 917 |
| Bioresmethrin | 8600 | 0.2 | (flies) | 43,000 |

*Source:* Ian J. Graham-Bryce, ''Crop Protection: Present Achievement and Future Challenge'' *Chemistry & Industry*, No. 13, 1976.

Which strategy or strategies to employ depends on the nature and cause of the exposure, the pesticides involved, and the resources available. Lowering dietary exposure, for instance, may require lengthening the time between spraying and harvesting the crop, thus allowing the pesticide to break down. Protecting field-workers, on the other hand, may entail changing safety procedures or using better equipment. Pesticide residues in food and water remain an important concern—especially where persistent, fat-soluble organochlorine chemicals are still in use—and one that can be answered mainly by eliminating or severely curtailing their use. But because most evidence suggests that occupational exposure is a more severe problem than dietary intake, the emphasis here is on ways to reduce risks to exposed workers rather than consumers.

## Assessing Occupational Exposure

Test data and experience show that many pesticides are dangerous, and intuition tells us that exposure can be reduced. Unfortunately, we do not know in most instances how to determine how much chemical is reaching workers or whether a given amount is too much. Hence, a major weakness in any effort to reduce exposure is the state of knowledge about existing exposure levels.[17] No general principles of pesticide exposure have emerged from research to guide the design and evaluation of studies. Although exposure assessment is receiving greater attention in academia, government, and industry,[18] the science remains largely empirical, and the goal of most studies is still primarily to answer regulatory concerns.

The methods for determining exposure levels vary from study to study, but the basic idea has changed little in thirty years. The "patch" method introduced by Batchelor and Smith in 1954[19] and popularized by Durham and Wolfe in the early 1960s[20] involves placing small pieces of gauze, chromatography paper, cloth, or other material on or under a worker's clothing. These patches are removed after the worker performs one or more tasks involving the pesticide, and the material is analyzed for pesticide residues. Total dermal exposure is then calculated by extrapolating from these small surface areas to average or "standard" body surface areas. However, researchers do not yet know whether the residues deposited on the patches are the same as those deposited on skin, or whether dosimeters accurately represent total dermal exposure.[21]

Comparing exposure studies is all the more complicated because scientists don't agree on what the patch should be made of and whether its backing should be permeable. Estimating inhalation exposure involves similar uncertainties. Often, inhalation is simulated using a small air pump with a filter attached, and then total exposure is calculated by extrapolating to assumed respiration rates. But since these rates will vary from worker to worker, room for error is considerable.

Other determinants of reliability in exposure studies include the number of dosimeters per worker, the location of the dosimeters, the number of workers tested, and the time that the dosimeter is in place. Overall, few of the 200 or so exposure studies to date have been statistically rigorous, so generalizations about pesticide exposure are hard to defend.

Faced with this uncertainty, researchers and policy-makers alike would like to fall back on "common knowledge" and "reasonable assumptions." But this approach does not always bear fruit. For example, although it is often assumed that rubber gloves provide 100 percent protection of the hands, recent studies show otherwise. Similarly, many scientists no longer believe that clothing necessarily protects covered skin fully. The general assumption that exposure is proportional to the amount of active ingredient used per acre and the number of acres treated has not been tested. While it seems logical that exposure in a closed tractor cab is less than in an open cab, the data are conflicting.

> **Overall, few of the 200 or so exposure studies to date have been statistically rigorous, so generalizations about pesticide exposure are hard to defend.**

Thus, our understanding of the dimensions of occupational exposure to pesticides is incomplete An extensive review of the literature on the subje revealed the following:

- In most studies, not enough data were reported to determine exposure rates or the factors that may have affected exposure.
- In many cases, the pesticide formulation and percentage of active ingredient were not reported. In some, the active ingredient was not even identified.
- Differences in conducting the studies and reporting values were so great that results from different studies cannot be reliably compared.
- The values reported for exposure were too heterogeneous to support quantitative conclusions.[22]

**On balance, given current knowledge, quantitative statements about applicator exposure—in relation to almost any factor—are well-nigh impossible.**

On balance, given current knowledge, quantitative statements about applicator exposure—in relation to almost any factor—are well-nigh impossible.

Can *no* conclusions be drawn about pesticide exposure? Certainly, just because the data base o exposure is so poor, the development and implementation of exposure-reduction strategies should not be postponed until better data can be collected: after all, human life is at stake.[23] Instead efforts to improve measurement methods and knowledge of exposure should go hand in hand with those to lower exposure, and emerging strategies should be evaluated on the basis of enhanced understanding. At a time when our knowledge is so limited, common sense, intuitior and analytical thinking still have a place in efforts to reduce exposure now.

### Exposure Reduction—A Starting Point

In the last two decades, Integrated Pest Management (IPM) programs have reduced pesticide use considerably on some crops. Typically, fewer applications per season on fewer acres are now needed, and sometimes dosages ar reduced as well.[24] But potential far exceeds practice. According to the Office of Technology Assessment of the United States Congress, pesticide use in U.S. agriculture could be reduced by as much as 75 percent if IPM were universally

**Pesticide use in U.S. agriculture could be reduced by as much as 75 percent if IPM were universally adopted. Even if exposure didn't fall proportionately, reductions would have to be substantial.**

percent if IPM were universally adopted.[25] Even if exposure didn't fall proportionately, reductions would have to be substantial since exposure among applicators and others working directly with pesticides is undoubtedly related to the number of times the chemicals are handled. As for fieldhands, it appears that with some pesticides the likelihood of acute poisoning increases with the total amount of chemical applied, so reducing use would also lower exposure to them.[26]

Protective clothing and other safety gear can also lower exposure to pesticides.[27] Indeed, with highly hazardous chemicals, such equipment is often mandatory where pesticides are carefully regulated. Considerable difficulties remain, however, with both its design and use. First, cumbersome and uncomfortable overgarments and equipment provide the best protection, and making sure workers use either requires extreme vigilance. In hot weather, some types of protective clothing may contribute to heat exhaustion.[28] Second, the effectiveness of some protective clothing is still at issue. While normal work clothing made of, say, denim offers some protection if its covers most of the body, the material itself can become contaminated and eventually *increase* exposure if not properly cleaned.[29] Furthermore, some pesticides are hard to remove from clothing and their residues may contaminate other laundry.[30] Some chemicals can permeate even rubberized or plastic-coated clothing.[31] Third, effective protective gear can cost so much that most potential users won't buy it save on pain of death.[32]

Regulation and user training are the two ways to make sure safety gear gets used. But even in the United States, where educational and enforcement institutions are relatively well developed, compliance with safety recommendations and regulations is low.[33] Training simply doesn't seem to improve compliance significantly. (For instance, one study of applicator-training programs found that participants *reported* greater use of rubber gloves during mixing and loading, but researchers could not verify an *actual* increase.)

Another major goal of applicator training is simply to get users to read pesticide labels and follow label instructions. Failure to follow directions on labels violates U.S. law, but labels contain so much information—some required and some intended to boost sales—that users can't

13

help being confused. (Manufacturers must be confused too: different products containing the same active ingredient have been found to have contradictory first-aid directions on their labels.) Moreover, depending on the user to read and follow instructions makes enforcement of pesticide safety largely a voluntary matter. Then, too, even in "advanced" countries, many people who handle pesticides may not be able to read.

Meanwhile, anecdotal reports abound of workers clad in little more than bathing suits while applying or handling pesticides. If safety is taken this casually in the United States, how can training and improved labels be expected to work in developing countries? Clearly, there must also be involuntary means of separating the person from the pesticide by changing the way chemicals are formulated, packaged, mixed, loaded, and applied.

## Pesticide Application Technology

How pesticide-application technology is developed depends on many factors, of which safety is only one. Cost, convenience, labor and time requirements, and efficacy are all high priorities, probably higher to most manufacturers than safety and environmental concerns.[34] Users' selection of equipment is even less likely to be influenced by safety considerations: one safety-conscious expert on pesticide application fails even to mention it among eleven criteria for choosing application equipment.[35] Small wonder, considering that so little research has been done on how application methods affect pesticide exposure levels.

Few types of application equipment have been tested for their effects on exposure, and very few of the many combinations of formulations and active ingredients have been tested with each type of equipment.[36] Surely, some differences among application methods exist. (See Table 4.) What decision-makers need is a set of guiding principles to use to improve safety in pesticide application until more extensive data are available.

Researchers do agree that "the rate of exposure per hour is greatest during measuring, mixing, and loading of a pesticide."[37] In studies of aerial application, mixers'/loaders' exposure to pesticides was estimated to be one hundred times higher than that experienced by spray pilots and ten times that of flaggers' exposure.[38] Since

*Anecdotal reports abound of workers clad in little more than bathing suits while applying or handling pesticides. If safety is taken this casually in the United States, how can training and improved labels be expected to work in developing countries?*

*What decision-makers need is a set of guiding principles to use to improve safety in pesticide application until more extensive data are available.*

**Table 4. Estimated Applicator Exposure from Different Equipment**

| Type of Equipment | Average exposure ±S.E.* | Range* | Number of Observations |
|---|---|---|---|
| Airblast | 790 ±440 | 109-2,826 | 283 |
| Handheld hydraulic sprayguns | 340 ±640 | 0.8-2,175 | 12 |
| Knapsack sprayers | 320 ±460 | 20-11,518 | 20 |
| Portable mistblowers | 150 ±50 | 19-546 | 6 |
| Hydraulic ground boom sprayers | 210 ±160 | 0.03-3,460 | 15 |

*Values are micrograms active ingredient per 100 square centimeters surface area per hour.
*Source:* Douglas G. Baugher, "Exposure to Pesticides During Application: A Critical Review of the State of the Art," consultant report to World Resources Institute, October 1984.

on-the-ground workers handle concentrated liquids or powders, spillage, splashing, or puffing pose especially great hazards during application. Only by physically separating the person from the pesticide can exposure be reduced or eliminated. Where possible, of course, the best method is simply not to mix or pour toxic chemicals at all. For instance, switching from a liquid to a granular formulation may be feasible. Granules can be applied with hand-held, knapsack- or tractor-mounted equipment, or by air.[39] Coating the granules with waxes, starches, or polymers can slow the release of chemical from the substrate of clay, sand, or other material. In addition, application rates for some agricultural uses can be reduced by placing granules in or near the seed furrow, creating a narrow band of protection very close to the plant and below ground, which lessens risk to people after the pesticide has been applied.[40] Another way to eliminate the mixing step in agricultural applications is to use seed that has been pretreated with pesticide. Several chemicals can be loaded onto the same seed (especially those that have systemic action) in carefully controlled facilities instead of the field.[41] Warning colors and bitter additives must, however, be added to deter wildlife and keep treated seed out of the human food chain.

Where only liquids will do, premixed formulations, preferably in loadable containers, can be used to reduce exposure.[42] For example, ICI's "Electrodyn" sprayer uses a glass bottle with its own built-in nozzle (the "Bozzle") in lieu of the spray tank. The bottle contains the ready-to-apply

15

pesticide.[43] Alternatively, mixing could take place automatically during spraying. In an "injection system," the concentrate and diluent (usually water) are loaded into separate containers on the spray rig, so no one is forced to pour a concentrated liquid. (See Figure 1.) This system has an additional advantage too: there is never any leftover diluted pesticide requiring disposal.[44]

If mixing cannot be avoided, exposure can be drastically reduced by using soluble packets—single-dose packages that dissolve in the sprayer tank[45]—or by using closed systems for transferring concentrate into the sprayer tank.[46] Most closed systems consist of a probe inserted into the pesticide container, a pump for transferring the concentrate into the spray tank, a meter for measuring the amount of concentrate transferred, and a shielded orifice where the concentrate enters the tank. However, simpler versions can be locally constructed or manufactured for use in either developing or technologically advanced countries. (See Figure 2.) In these approaches, no concentrate is poured from one container to another, so spilling or splashing is minimal. Since California imposed closed-system requirements for certain highly toxic pesticides, the number of poisoning incidents among mixers and loaders has declined. (See Table 5.) Although other factors have influenced this trend, closed systems have no doubt helped reduce the poisoning rate.[47]

Drift—the movement of pesticides through the air from the intended site of application—can be a major source of pesticide exposure for those using the chemical and for field workers, nearby residents, and wildlife. Depending on a host of variables, pesticide can drift considerable distances, putting even larger populations at risk. Since drift means waste, it adds unnecessarily to farmers' costs too. Thus, making application more efficient benefits the user even as it safeguards "nontarget populations" from unnecessary exposure—a "win/win" game.[48] In addition, drift of herbicides onto other crops can lead to enormous economic losses: one scientist has estimated that drift caused $11 million in crop damage in Texas alone last year.[49] Clearly, control of drift deserves a high priority for economic as well as safety reasons.

One way to reduce drift is to apply pesticides in heavier particles that fall to the ground instead of waft through the air. Since granules are many

**Making application more efficient benefits the user even as it safeguards "nontarget populations" from unnecessary exposure—a "win/win" game.**

**Figure 1: Schematic of Injection System**

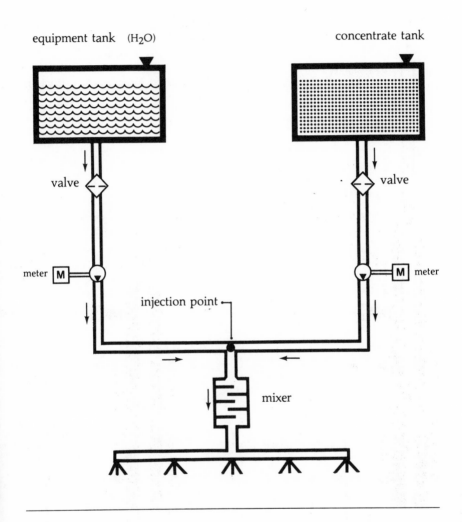

equipment tank (H₂O)

concentrate tank

valve

valve

meter

meter

injection point

mixer

*Source:* Schmidt, "The Direct Injection Technique for Preparing the Spray Mix"

times heavier than liquid droplets of the same size, they are often the answer.[50] Where applying a liquid pesticide is essential, the formulation can be modified to assure that droplets remain large enough to minimize drift. Spray nozzles designed for both aircraft and ground equipment can be

17

# Figure 2: Example of Closed System:
## Vacuum Probe Transfer and Rinse System

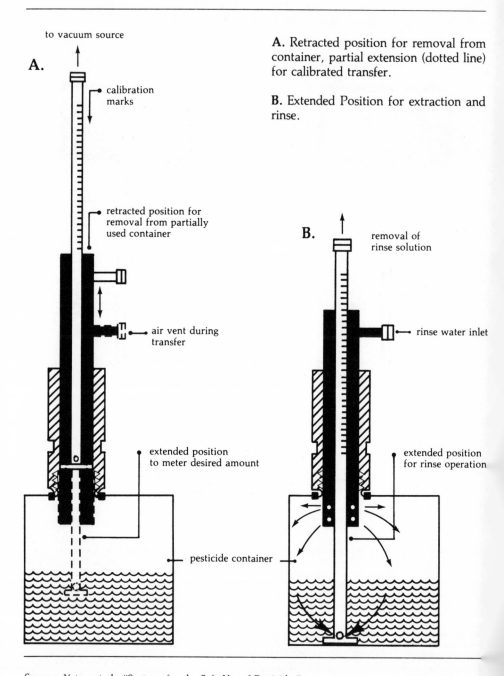

**A.** Retracted position for removal from container, partial extension (dotted line) for calibrated transfer.

**B.** Extended Position for extraction and rinse.

to vacuum source

**A.**

calibration marks

retracted position for removal from partially used container

air vent during transfer

extended position to meter desired amount

**B.**

removal of rinse solution

rinse water inlet

extended position for rinse operation

pesticide container

*Source:* Yates, *et.al.,* "Systems for the Safe Use of Pesticides"

# Figure 2: Examples of Closed Systems (continued)

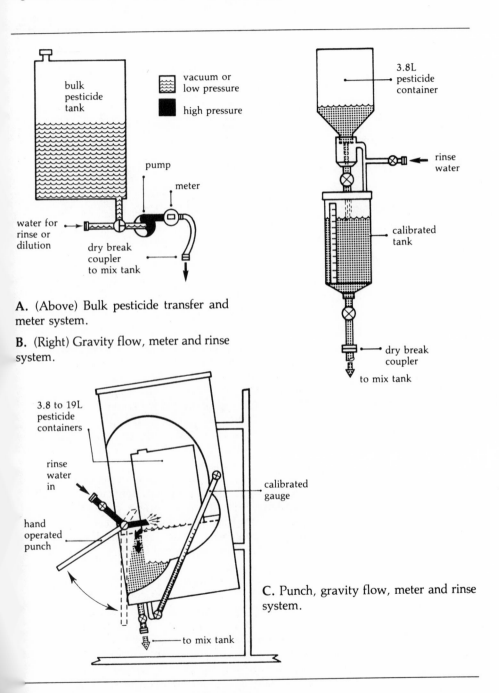

A. (Above) Bulk pesticide transfer and meter system.

B. (Right) Gravity flow, meter and rinse system.

C. Punch, gravity flow, meter and rinse system.

Source: Yates, et.al., "Systems for the Safe Use of Pesticides"

# Table 5. Employees Reported to Have Pesticide Illnesses in California

| | Total numbers by employment | | | | | | |
|---|---|---|---|---|---|---|---|
| | 1973 | 1974 | 1975 | 1976 | 1977 | 1978 | 1979 |
| Pilots | 14 | 17 | 7 | 8 | 7 | 8 | 1 |
| Flaggers | 20 | 6 | 16 | 14 | 15 | 24 | 9 |
| Mixer/loaders | 165 | 141 | 143 | 122 | 143 | 142 | 132 |
| Agr. ground applicators | 424 | 225 | 264 | 254 | 236 | 163 | 161 |
| Field worker re-entry | 157 | 112 | 167 | 156 | 184 | 95 | 53 |
| Gardeners (commercial) | 66 | 103 | 106 | 159 | 155 | 138 | 92 |
| Nursery and greenhouse | 112 | 73 | 90 | 119 | 72 | 69 | 48 |
| Fumigator (field) | 71 | 29 | 22 | 14 | 16 | 20 | 21 |
| Machine clean and repair | 22 | 28 | 40 | 26 | 33 | 21 | 19 |
| Tractor drivers, irrigators | — | 23 | 22 | 30 | 29 | 20 | 23 |
| Drift exposure (persons nearby) | 26 | 22 | 31 | 29 | 30 | 44 | 19 |
| Other* | | | | | | | 30 |
| Total all of agriculture | 1077 | 779 | 908 | 931 | 920 | 744 | 608 |
| Total other than agriculture | 397 | 378 | 435 | 521 | 598 | 450 | 411 |
| Grand total for year | 1474 | 1157 | 1343 | 1452 | 1518 | 1194 | 1019 |
| % by agriculture | 73 | 67 | 68 | 64 | 61 | 62 | 60 |

*1979 agriculture includes, for the first time, animal applicator and self-employed categories.
Source: Physicians' First Report of Illness and other medical reports, California Departments of Food & Agriculture, Industrial Relations, and Health Services.

used to enlarge droplets, too. (See Figure 3.) In addition, application equipment can also be modified to reduce drift: for instance, "shrouding" the spray booms of ground equipment keeps droplets from swirling up into the air and makes it possible to apply pesticides when it's windy.[51] And on the drawing board are aircraft modifications that reduce drift during aerial application.[52]

Unfortunately, large droplets of pesticides are not usually as effective as smaller ones, especially on plants. With spotty coverage, protection may decline, requiring re-application or causing unacceptable losses. Also, large droplets are more likely to land on the ground, contaminating the soil unnecessarily.

Given these drawbacks, a more unconventional way to reduce drift and inefficiency—improving deposition by electrostatically charging pesticide droplets—is worth a closer look. In this method, which keeps droplet size uniformly small, deposition on the target is maximized. A hand-held electrostatic sprayer is already on the market, principally in developing countries, while a tractor-mounted unit has yet to be commercialized

## Figure 3: Schematics of Low-Drift Nozzles

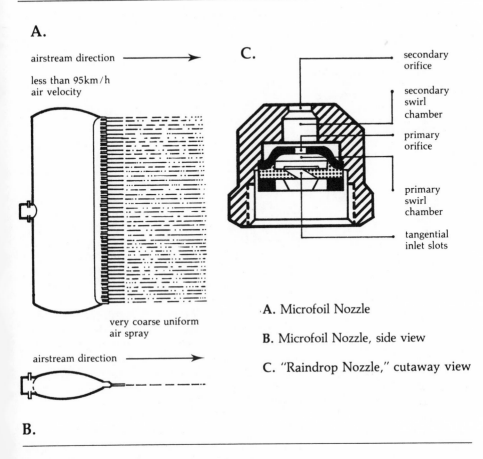

**A.**

airstream direction ⟶

less than 95km/h
air velocity

very coarse uniform
air spray

airstream direction ⟶

**B.**

**C.**

secondary orifice

secondary swirl chamber

primary orifice

primary swirl chamber

tangential inlet slots

A. Microfoil Nozzle

B. Microfoil Nozzle, side view

C. "Raindrop Nozzle," cutaway view

*Source:* Matthews, *Pesticide Application Methods*

The evidence so far suggests that, using either, farmers can get by with applying substantially less pesticide per acre with no loss of effectiveness.[53] More extensive studies of applicator exposure using electrostatic sprayers are needed, but preliminary results indicate that such equipment improves safety too. (See Table 6).

Often, disposing of the pesticide container, unused diluted pesticide, and rinsate from cleaning the container or the equipment poses risks to the user and the equipment. Most

21

**Table 6. Applicator Exposure Using Conventional and Electrostatic Sprayers**

| Country | Device | g active applied per hour | Formulation | Crop height cm | TDC* mg/hr | TDC* as % of active ingredient applied |
|---------|--------|---------------------------|-------------|----------------|------------|----------------------------------------|
| Tanzania | Electrodyn sprayer | 16 | Cypermethrin | 30-60 | 26.9 | 0.17 |
| | Spinning disc | 116 | Cypermethrin | 30-60 | 369.9 | 0.32 |
| Ivory Coast | Electrodyn sprayer | 21.6 | Cypermethrin | 120-160 | 8.9 | 0.04 |
| | Spinning disc | 23.4 | Cypermethrin | 110-180 | 17.8 | 0.08 |
| Paraguay | Electrodyn sprayer | 6.6 | Cypermethrin | 66-125 | 3.0 | 0.05 |
| | Knapsack | 13.1 | Cypermethrin | 75-180 | 29.5 | 0.22 |

*TDC—Total dermal contamination
Source: T.B. Hart, "The Hand-Held 'Electrodyn' Sprayer: Worker Hazard," ICI Plant Protection Division (Fernhurst, England: undated).

dangerous is using pesticide containers to store or carry food or water.[54] One way to solve this problem is to use soluble packets. Another is to design containers like the "Bozzle" that cannot be easily re-used.[55] Using these alternatives can also obviate the problem of storing partially-used containers. Where dry formulations such as granules or treated seeds are used, the bags can sometimes be burned or buried with few special precautions. Pesticide-injection systems can be automatically flushed, spraying the rinsate over the crop during the last few minutes of operation. Granule applicators using an air stream can be flushed in much the same way.[56]

Most pesticide application is inefficient largely because the chemical must be broadcast into the environment so as not to miss the pest. Methods that reach the pest better or that attract the pest to the pesticide can thus help reduce the total amount of chemical needed and lessen its effects on nontarget organisms. Using treated seeds and carefully placed pesticide granules, leaves all but a small zone around the growing plant untouched by chemicals. Electrostatic sprayers mainly coat the undersides and edges of leaves with pesticide,[57] increasing the chances that certain leaf-eating insects will contact the chemical. Pesticide-loaded baits containing either a food source, a feeding stimulant, or some other attractant (such as sex pheromones) can be used to lure pests to the chemical, thus considerably reducing the amount of pesticide needed to do the job.

The developing technology of "controlled release" holds great promise for the more careful management of pesticides. If protected against premature breakdown in the environment, chemicals can be applied in minute quantities and released slowly, prolonging their effects. Pheromones for insect control and monitoring are already "slow-released" in synthetic fibers and capsules, and toxicants could be too.

This brief review of improved application and formulation technologies indicates the range of options available for enhancing human and environmental safety in pesticides use. Many can be used now. Others only need fine tuning to be more widely utilized. Why, then, aren't such techniques being implemented? Why does safety remain such a low priority in technology development and transfer? Who is responsible for moving these methods forward, and under what constraints? What can be done to assure that safer means of pesticide application are devised, developed, and used?

## Policy Implications of Safer Application

Pesticide application can be made significantly safer only if the chemical industry, the farm-equipment industry, government research and regulatory agencies, and university researchers and extension specialists are all involved. While each of these actors has its own goals, objectives, and obstacles, interaction among these sectors is critical to success.

The chemical industry's concerns are discovering, developing, and marketing pesticides. Pesticide companies control the formulation of active ingredients into marketed products, testing myriad solvents and carriers before releasing a final product. These formulations must meet a host of requirements, such as efficacy and stability in a broad range of environments. Significantly, formulations must usually be "designed to perform in the readily available equipment used at customary operating conditions."[58] This is understandable: companies seeking to recoup sizable investments and make a profit don't want to restrict products to owners of new, specialized equipment. Meanwhile, equipment manufacturers want to maximize sales so they design equipment to make use of available chemical formulations. Who should change first, and what's in it for them?

**Companies seeking to recoup sizable investments and make a profit don't want to restrict products to owners of new, specialized equipment.**

23

Interestingly, the major exception to this conundrum is the Electrodyn/Bozzle, which uses a special mineral-oil formulation. The Electrodyn is manufactured by the equipment division of a major chemical company, ICI, which can sell an integrated package of applicator and chemical. (Few other companies have both chemical and equipment divisions.) ICI markets pesticides formulated for use in the Electrodyn. So far, the device has been marketed only as a hand-held sprayer for use in developing countries.

Larger-scale electrostatic sprayers are not available partly because of technical difficulties and partly because of marketing considerations. Another electrostatic sprayer, developed at the University of Georgia, has yet to be commercialized in the United States despite the promising performance of prototypes. The first licensee, FMC (like ICI, a chemical company with an equipment division) decided not to complete development of the sprayer. The market reality is that selling chemicals is more profitable than selling sprayers and electrostatic sprayers apply significantly less pesticide than the models they would replace. In fact, according to the Food and Agriculture Organization of the United Nations, for every research dollar spent on developing application technology in 1977, $4,000 was spent discovering and developing new chemicals.[59]

Pesticide application equipment comprises only a sliver of a relatively small and fragmented farm-equipment industry in the United States. Little capital is available to companies for research and development,[60] while the market for new products is uncertain and resistant to change.[61] Application equipment, a capital expense for the user, often lasts 15 to 20 years, and so will be replaced only when necessary. While some users, most notably growers of such high-value crops as tree fruits, acknowledge the need to improve application efficacy and efficiency,[62] they have little incentive to invest in new equipment before their old equipment wears out.

In short, manufacturers of pesticides and equipment have no strong reasons to change the status quo. The current chemicals and formulations are designed to be applied with equipment already widely used, the equipment is designed to apply readily available chemical formulations, and for most users these systems work. Even if application is inefficient and safety could be improved, why should

**Manufacturers of pesticides and equipment have no strong reasons to change the status quo.**

any one of these three groups change the way it does business? Who should take the first steps?

So far, the U.S. government has demonstrated only a questionable commitment to safer pesticide application. In the United States, from 1968 to 1978, the number of engineers in Land Grant universities working on pesticide application declined 37 percent, from 35 to 22. In terms of time devoted to application technology, the decline was more precipitous: scientist-years dropped 43 percent.[63] Clearly, administrators of publicly funded research don't see this as a high-priority area.

> If market forces cannot bring new technologies to the fore, perhaps government intervention can.

But government interest is a necessity: If market forces cannot bring new technologies to the fore, perhaps government intervention can. California's experience with requiring closed systems, offers a useful example. Moreover, much can be learned from U.S. federal standards for automotive fuel efficiency and emissions. Despite obvious differences between the auto and farm-equipment industries, setting efficiency standards for application equipment could—like those for vehicles—work if they were accompanied with appropriate incentives. One possibility would be for federal or state agencies with sizable pesticide-application responsibilities (among them, the Forest Service, the Bureau of Land Management, or the armed services) to develop specifications for new, more efficient, and safer application equipment, and to request designs and bids for large equipment orders. Given the potential for a large and virtually guaranteed market, companies might be more willing to invest the necessary R&D funds to develop new ideas. If the contracts were big enough to pay back the original investment, such companies would probably develop similar models for other markets.

Without such initiatives, the chemical and equipment industries may find themselves confronted with unwanted and potentially disruptive regulations. According to a representative of the National Agricultural Chemicals Association, at least 15 states are considering legislation on "chemical trespass," defining liability for damages due to pesticide drift. In the United Kingdom, legislation has been introduced that would require pesticide label instructions specifying the types of equipment that can be used in applying each product.[64] Such laws could bring about major and unanticipated changes in pesticide use, determining which chemicals are applied where, which crops are grown where, and who applies the pesticides. These changes could unleash chaos if no new technologies are available to meet users' new needs. Yet, if industry and government take up the issue of pesticide application now, the technology could evolve rationally.

## Application Technology and Improved Pest Management

Effective pest management requires that chemicals be used only when and in the amounts needed. Pesticides that don't reach their intended target represent unnecessary costs, and they can wipe out pests' natural enemies. Unfortunately, some application methods may work directly against new pest-management approaches. For instance, the lack of an efficient means to apply microbial pesticides limits their economic viability. So too, dependence on aerial applications, booked long in advance instead of applied as needed, makes it difficult to time and place pesticide treatments to have the optimal effect. And spot treating some areas while leaving others unsprayed—a tactic that may help manage resistance[65]—is hard with equipment designed to blanket large areas with chemicals. In general, using more efficient ground equipment might reduce reliance on aerial spraying, thus reducing waste and risks.

The safest and most effective pest-management programs must take into account not only the effects of the pesticides and other "inputs," but also the way they are applied. Environmental contamination, human exposure to pesticides, and efficacy are linked in complex ways, and application methods need to be developed and used with all three in mind. With new chemical and microbial pesticides, efficiency and safety are all the more important. If either is ignored, many of the possible benefits of better pest management may go unrealized.

# III. Putting Nature to Work: Biological Methods of Pest Control

Most projections of future use of pest-management technologies are based on the assumption that chemical pesticides will be the principal pest-control method at least until the next century.[66] But the need for chemicals may well be overstated. Do pesticide-use levels reflect the inherent superiority of chemical controls over other methods, or do they indicate the amount of research dedicated to developing chemicals instead of alternative controls?

This critical question has no straightforward answer. Dependence on pesticides varies among cropping systems and regions, and so do the expected effects of withdrawing chemicals from these systems. Despite these differences, however, the U.S. Congressional Office of Technology Assessment still concluded that the narrow range of control options available to U.S. farmers poses a major constraint to improving crop protection.[67] Then too, pesticides have been used in commercial agriculture for so long that they have greatly modified the ecological structure of many agroecosystems so that drastic losses due to pests often occur when chemicals are removed.[68] Indeed, only carefully redesigned cropping systems can demonstrate the actual differences between chemical-intensive and alternative approaches. Yet, while the potential for replacing chemicals with alternatives cannot be quantified, nonchemical methods could at least partially substitute for pesticide-based management—

27

especially in developing countries, where comparatively less capital and energy are available for agriculture.

For comparing chemical controls with their alternatives, agreed-upon criteria are needed. Since agricultural productivity must continue to increase simply to keep pace with demand, alternative pest-management methods must be as reliable and effective as conventional approaches. But whether these new methods need to match chemicals in reducing pest-caused losses is debatable because the stability (or long-term efficacy of pest-management technologies from year to year) and sustainability (their effectiveness over time) are also important. Another concern is equity—assuring that the risks, costs, and benefits of pest-management technologies are not unfairly distributed between rich and poor, or between large and small landholders.[69] Such criteria must eventually be used to evaluate the system-wide effects of all pest-management technologies. So far, chemical pesticides have been incorporated piecemeal or wholesale into agriculture and elsewhere without full regard for the consequences. But with the "biotechnology" revolution upon us, the time for grasping the consequences of coming innovations may be short indeed.

## Biological Control

In undisturbed natural environments, outbreaks of plant-destroying insects and diseases are relatively rare. Among other forces, a host of natural enemies and competitors keep the numbers of insects, microbes, vertebrates, and plants in check. In agriculture and other managed ecosystems, natural enemies—predators, parasites, and pathogens—have been used for decades to manage pest populations. For weed control, plant-feeding insects are also part of the armamentarium.[70] To control fungi and bacteria that cause plant diseases, "antagonists"—other microbes that either destroy or exclude pathogens from infection sites on the plant—can be used.[71]

The three approaches to biological control are inoculative release ("classical" biological control), inundative (or augmentative) release, and the conservation of existing populations ("natural" control). Modern biological control by inoculative release began when the vedalia lady bettle was

**The three approaches to biological control are inoculative release ("classical" biological control), inundative (or augmentative) release, and the conservation of existing populations ("natural" control).**

imported into California in 1888 to control cottony-cushion scale on citrus.[72] Although it has made occasional comebacks when insecticides were overused, this pest—which threatened to wipe out California citrus in the 1880s—has remained innocuous to this day, without resort to chemicals. This spectacular success triggered research and introduction programs in many parts of the world. In Australia, the technique was used to control the prickly-pear cactus, which had infested 24 million hectares of rangeland and rendered half of them useless for grazing. Introducing a host-specific moth in 1925 reduced this range weed to levels below those that would cause economic loss. Similarly, a virus was used to control jackrabbits that were damaging Australia's rangelands. As of 1976, some 327 cases of biological control based on the introduction of natural enemies had met with at least partial success. (See Table 7.)

Permanence is one advantage of introducing natural enemies in this way. Thus, the payoffs in future crop protection and forgone chemical costs are virtually limitless. For example, it cost about $750,000 to find and introduce *Chrysolina* beetles in the 1940s to control a toxic range weed, *Hypericum perforatum*, in California. Savings to date have been estimated at more than $100 million.[73] Other estimates of "benefit/cost ratios" for biocontrol range well over 100:1. (See Table 8.)

**Introducing new species into a habitat entails some risk that what was brought in to control pests might become a pest itself.**

Introducing new species into a habitat entails some risk that what was brought in to control pests might become a pest itself. Were this to occur, environmental costs would have to be considered in calculating the net benefits of biological control. Because of the extreme caution that has been taken, however, modern biological control specialists can boast that the environmental cost of this approach is virtually zero.[74] By comparison, the environmental costs of chemical pesticides have led researchers to reduce benefit/cost estimates for chemical control by 25 percent, from 4:1 to 3:1.[75]

Classical biological control has a remarkably high success rate—55 to 75 percent[76]—mainly because those pests most likely to be controlled biologically are targeted. Since most of these target pests are themselves introduced species, researchers can efficiently study and search the pest's original habitat for biocontrol agents. Also, most pests chosen for this line of attack are perennial weeds or pests of perennial crops, so

# Table 7. Biological Control Successes

### Biological Control of Insect Pests
### (Successes listed for only one geographical location
### for each species—where first introduced)

| Degree of Success* | Islands | Continents | Total |
|---|---|---|---|
| Complete | 12 | 19 | 31 |
| Substantial | 32 | 41 | 73 |
| Partial | 25 | 28 | 53 |
| Total | 69 | 88 | 157 |

### Ratings of Biological Control of Insect Pests
### (Successes listed for every location into which importation has occurred)

| Degree of Success* | Islands | Continents | Total |
|---|---|---|---|
| Complete | 35 | 67 | 102 |
| Substantial | 58 | 86 | 144 |
| Partial | 36 | 45 | 81 |
| Total | 129 | 198 | 327 |

*Complete successes refer to complete biological control maintained against a pest of a major crop over an extensive area so insecticidal treatment is rarely, if ever, necessary. Substantial successes indicate that economic savings are less pronounced because (a) the pest or crop is less important, (b) the crop area is limited, or (c) occasional insecticide applications are needed to maintain control. Partial successes demonstrate that chemical control remains necessary, but either the intervals between applications are lengthened, or the results are improved when the same treatments are used, or outbreaks occur less frequently.

Source: F.W. Stehr, "Parasitoids and Predators in Pest Management," in R.L. Metcalf and W.H. Luckmann, eds., *Introduction to Insect Pest Management* (New York: John Wiley and Sons, 1982), pp. 135–73.

permanently establishing the natural enemies in the environment is relatively easy. Without diminishing the importance of the success rate of this approach, these observations indicate the value of conducting research in support of biological control. Indeed, a century's worth of evidence shows plainly that the degree of interest and effort is the main determinant of biocontrol's success.[77] New analyses suggest, too, that strategies of searching for natural enemies could be modified to improve the chances of success.[78]

Biocontrol agents need not become a permanent part of the ecosystem to be effective. Inundative or augmentative release involves rearing and periodically releasing large numbers of natural enemies to temporarily suppress pest populations. Beneficial species are used as living pesticides and applied as needed to bring pest numbers down to tolerable levels. Unlike classical biological

# Table 8. Some Estimates of Benefits From Biocontrol

## A. California 1923–59

| Pest (date controlled) | Total savings over previous losses plus pest control costs (US$ millions) |
|---|---|
| Citophilus mealybug (1930) | 56.0 |
| Black scale (1940) | 32.0 |
| Klamath weed (1953) | 20.9 |
| Grape leaf skeletoniser (1949) | 0.75 |
| Spotted alfalfa aphid (1958) | 5.6 |
| Total savings on 5 projects | 115.3 |
| Total spent over period | 4.3 |

## B. Australia projected to year 2,000 at 1975 values discounted 10%

| Pest | Cost of research by CSIRO* (Australian$) (in millions) | (U.S.$) | Total benefits (Australian$) (in millions) | (U.S.$) |
|---|---|---|---|---|
| Skeleton weed | 2.39 | 3.13 | 264.4 | 346.41 |
| Wood Wasp | 6.27 | 8.21 | 15.4 | 20.17 |
| White Wax scale | 1.07 | 1.40 | 1.7 | 2.22 |
| Two-spotted mite | 0.67 | 0.88 | 17.5 | 22.92 |
| Red Scale | not available | | 17.7 | 23.19 |

## C. Benefits of some tropical CIBC** projects as supplied by sponsors

| Pest | Country | Total Cost US $ | Benefit (year assessed) US $ |
|---|---|---|---|
| Rufous scale | Peru | 1,789.14 | $226,916 per year saved on chemical control (1977) |
| Barnacle scale | Hawaii | — | $110,000 per year increased yield (1974) |
| Sugarcane scale | Tanzania | 11,975.88 | $67,832.16 per year increased yield (1974) |
| Sugarcane borer | Brazil | 9.54/ha p.a. | $48.15/ha per year (1979) |
| Coconut leafminer | Sri Lanka | 35,782.95 | $12.4 million per year in crop saved (1977) |
| Potato tuber moth | Zambia | 35,731.66 | $547,559.78 gain discounted from 1974 to 1980 (1974) |

*Commonwealth Scientific Industrial Research Organization
**Commonwealth Institute for Biological Control
Source: D.J. Greathead and J.K. Waage, *Opportunities for Biological Control of Agricultural Pests in Developing Countries* (Washington, D.C.: The World Bank, 1983).

control—necessarily a public function because its benefits are widely shared—inundative biocontrol can be commercialized. Several microbes, most notably the insect pathogen *Bacillus thuringiensis* (B.t.), have been registered and used as pesticides in many parts of the world.[79] B.t. has been widely

*example:* used to control various species of caterpillars, and a recently discovered variant has been found effective against mosquitoes and blackflies.[80] In the United States, a protozoan pathogen of grasshoppers has been used to combat the outbreaks of these pests in the Great Plains.[81] Microbes can also be used to control some weeds[82] and plant diseases.[83] Beneficial insects and mites are available from commercial producers in the United States[84] and Europe, and farmers' cooperatives have set up their own rearing facilities in Peru,[85] China,[86] and elsewhere. In Europe, the greatest success with mass releases of "beneficials" has been in the control of whiteflies and spider mites in commercial greenhouses.[87] In the United States, soybean farmers effectively and economically control the Mexican bean beetle by annually releasing a parasitic wasp that the U.S. Department of Agriculture supplies at cost.[88]

Like chemical pesticides, these beneficial agents must be produced, maintained, stored, transported, and applied—which, unlike classical biological control, entails continuing costs. Still, this approach can be cost-effective,[89] especially where high-value crops such as tree fruit or greenhouse ornamentals and vegetables are at stake.[90] Where the risks of human exposure to chemicals need to be eliminated, or sensitive natural habitats protected, using a microbe as a pesticide instead of a chemical may be justified if social costs are considered along with simple material and operational costs of using chemicals. For example, *Bacillus thuringiensis* is sometimes used to control gypsy moth and spruce budworm in the forests of the northeastern United States in or near inhabited areas.

Using electrostatic sprayers or other more efficient application methods may make microbial pesticides more effective. In addition, new techniques in biotechnology—especially genetic engineering and tissue culture—may open avenues for the wider use of this type of biocontrol. Already, the Monsanto Corporation has announced a genetically-engineered bacterium that it proposes to register as a seed treatment for corn rootworm control, probably the biggest insecticide market in the United States. Other biotechnology companies are looking at ways to broaden the number of insect species that *Bacillus thuringiensis* can control. The bollworm/budworm complex in cotton, insects that damage fruits and vegetables, and such forest pests as the gypsy moth and spruce budworm are ready targets of insect viruses, several of which are now registered pesticides. Research to reduce variability in performance and improve propagation techniques will likely make these viruses more usable,[91] and biotechnological breakthroughs may help make the difference between commercial success and failure.

By far the most common form of biological control practiced is "natural control"—the conservation of natural enemies by preventing their destruction or preserving their habitats.[92] Selection and timing of pesticides, choice of plant varieties, maintenance of alternative hosts, and proper soil management are among the tactics employed to keep beneficial species active and populous enough to control pests. Control of mites in apples,[93] budworms and bollworms on cotton,[94] alfalfa weevil in alfalfa,[95] and various soil-borne pathogens in greenhouse crops is greatly enhanced by these methods, even though pesticides still form the backbone of programs to manage these pests. In addition, according to recent U.S. Department of Agriculture studies, such organic-farming methods as crop rotation, mulches, and intercropping—principally meant to improve the soil—also enhance pest management by protecting beneficial species.[96] Some evidence also suggests that increasing crop diversity through intercropping or polyculture reduces damage from insect pests by providing habitat for natural enemies.[97]

Natural pest control needs more work and could do more. Paradoxically, the future of natural control depends in part on developments in pesticide chemistry. If chemical insecticides were targeted more specifically at key pests, more predators and parasites would survive to suppress secondary pests. For example, chemicals might be designed that beneficial species can detoxify but plant-feeders cannot.[98] (Here, perhaps, biotechnological means could provide the complex molecules needed.) The same holds true for fungicides, some of which destroy naturally-

**Selection and timing of pesticides, choice of plant varieties, maintenance of alternative hosts and proper soil management are among the tactics employed to keep beneficial species active and populous enough to control pests.**

occurring fungi that could help control plant pathogens and insect pests. Another promising approach in natural control has been demonstrated in California and Michigan: pesticide-resistant predatory mites have been bred in the laboratory and released into orchards, where they multiply in the face of continued chemical application aimed at pests that cannot now be controlled biologically. These resistant predators effectively control mites that otherwise become major pests when chemicals kill off nonresistant beneficials.[99] Still unexplored is the applicability of this tactic to other pest problems. In general, progress on natural control depends on more studies of pesticide-free agroecosystems to determine how best to use and protect potential biocontrol agents.[100]

Biological control's full potential to save on chemical costs, increase yields, and enhance environmental quality has yet to be realized. In the poorest countries, where precious foreign exchange is being spent on the dubious short-term gains of some chemical-based control programs (often at the expense of long-term public health), this unrealized potential seems particularly unfortunate. Plantation crops, orchards, pastures, and rangeland are especially ripe for classical biocontrol in these countries.[101] Moreover, the inundative release of beneficial insects and other organisms may be far more adaptable to small-scale operations in developing countries than it has been in industrialized nations. Many of the rearing methods are well known and few are patented, so farmer cooperatives, local government units, or small businesses would need little start-up capital to produce them. Technical assistance on quality control in propagation, timing of releases, and selection of compatible pesticides may be needed to assure efficacy. As biotechnological methods of gene transfer and mass propagation increase the potential for microbial pesticides, new options for Third World agriculture may open up since microbes can be produced in smaller, simpler facilities than those needed to manufacture chemicals. Local manufacture means shorter storage time too, so more products could economically be made available through licensing agreements.

While biological control cannot solve all pest problems, pests' natural enemies exist in virtually

> As biotechnological methods of gene transfer and mass propagation increase the potential for microbial pesticides, new options for Third World agriculture may open up since microbes can be produced in smaller, simpler facilities than those needed to manufacture chemicals.

all ecosystems. How much we use them will be determined primarily by how much effort we expend to identify and understand their roles. A leader in the field has written that ''As long as we ignore, anywhere in the world, an effective natural enemy capable of controlling one of our major pests, we are postponing cheap, reliable and permanent control.''[102] We may also be sacrificing public health and environmental quality every time we opt for a potentially hazardous chemical when a benign biological control could be used as a partial or complete substitute.

## Host Resistance

The evolutionary history of animals and plants is in part one of attack and defense.[103] Some way has developed for warding off almost every insect or pathogen that has adapted to living off a plant or animal. And in the competition for water and nutrients, many plants have evolved chemical mechanisms to prevent others from occupying nearby space.[104] However unwittingly, farmers have selected these defenses—collectively called host resistance—from the time plants and animals were first domesticated.[105]

In modern times, pest-resistant varieties have been used most often to protect against plant diseases and insects. Today, as much as 75 percent of U.S. cropland is planted to disease-resistant strains, creating a net benefit exceeding a billion dollars.[106] In Asia, resistance bred into rice varieties helps account for the dramatic productivity gains over the last two decades.[107] According to the U.S. Council for Environmental Quality, ''From the standpoint of the farmer, pest-resistant varieties are usually the most effective, easiest, and most economical means of controlling insect and plant disease pests.''[108] For many pests, especially viruses attacking plants, there are no known alternatives to breeding for resistance. Overall, the ''benefit/cost ratio'' for research in host-plant resistance is about 300:1, excluding reduced pesticide costs.[109]

While host resistance has usually completely substituted for other forms of pest control, it is also effective as a partial replacement. Breeding can also create plants whose planting, maturation, or harvest don't coincide with pests' natural cycles. The use of short-season cotton has, for example,

**Overall, the ''benefit/ cost ratio'' for research in host-plant resistance is about 300:1, excluding reduced pesticide costs.**

done as much as any other control method to keep insects in check.[110] Resistance can decrease pest survival on the plant so that fewer insects or pathogens remain to cause damage, or it can slow down pest development so that populations don't peak during the growing season. (See Figure 4.)

---

Figure 4: Graphical Representation of Host Resistance

---

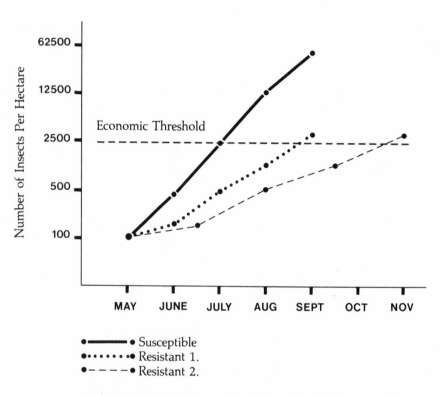

●━━━━● Susceptible
●●●●●●● Resistant 1.
●─ ─ ─ ─● Resistant 2.

Theoretical population trends of a hypothetical insect population on a (1)susceptible variety, (2)resistant variety 1 which reduces the pest insect population size by 50 percent in each generation and (3)resistant variety 2 which reduces pest population size by 50 percent in each generation and increases the duration of the developmental periods of the survivors by 50 percent. Assume five-fold increase in pest numbers per generation.

---

*Source:* Adkisson and Dyck, "Resistant Varieties in Pest Management Systems"

Even if protection runs out before the season ends, resistance may make it possible to delay the application of pesticides, thus reducing the destruction of natural enemies.[111] Used with judiciously applied pesticides, host resistance can thwart pesticide resistance in pest populations and limit pests' ability to overcome plant defenses.

Despite this tremendous potential, the future of host resistance remains uncertain because the diversity and accessibility of genetic material for breeding and the speed at which pests evolve ways to get around plant resistance are uncertain. Then too, where agricultural production becomes year-round instead of seasonal, or irrigation and fertilizer use increase, new demands on resistant crops and new opportunities for pests are created.

**And as the most productive countries and regions devote more and more land to large plantings of major crops, many of them genetic monocultures, global supplies could grow more vulnerable to new strains of pests, duplicating the serious losses that southern corn leaf blight inflicted on the U.S. Corn Belt in 1970.**

And as the most productive countries and regions devote more and more land to large plantings of major crops, many of them genetic monocultures, global supplies could grow more vulnerable to new strains of pests, duplicating the serious losses that southern corn leaf blight inflicted on the U.S. Corn Belt in 1970.[112]

In response to concern over the supply and conservation of genetic material, many valuable germplasm collections have been established here and abroad. But while these collections continue to expand, the characteristics—including resistance to pests—of the hundreds of thousands of strains they contain remain largely unknown. For the most part, the only way to identify traits is to raise plants and subject them to "challenge" by the pest, a time-consuming and labor-intensive process. And germplasm preservation has been limited to a few major crops, while many important but economically secondary species are virtually ignored.[113]

Besides gene banks, the other linchpin of genetic preservation is the deployment of genes in the environment. The Irish potato famine of the 19th century, an outbreak of coffee rust in the early 1900s, and the southern corn leaf blight epidemic of 1970 occurred because large areas were planted to similar genetic types. Once the pathogen evolved ways to overcome the plant's resistance in such "genetic lawns," virtually the entire crop lay open to infection, so losses were crippling. One promising defense against such disasters is planting mixtures of similar genetic types that differ mainly in their resistance characteristics.[114] Such "multiline" mixtures can halve the disease rate in some crops.[115] On a larger, regional scale, genetic diversity could be enhanced by maintaining some diversity of crops; that way, pests face inhospitable circumstances in which to grow and spread. Such tactics point up a generally neglected side of host resistance. If host resistance is to remain a viable strategy, plant breeders and pest-management specialists must begin to consider not only how effective resistance characteristics of plant types are. They will also have to ask how durable those characteristics will be in space and time,[116] departing from the current tendency simply to try and stay one step ahead of pests in breeding programs.[117]

Who will be doing this breeding? According to one scientist,

*The trend is inexorably toward commercialization of all plant breeding, plus contributions from a few major international institutes with a few . . . crops. A refreshing aspect of disease resistance breeding has been its historical development in universities, with their urgency to publish and share results, and to determine the mechanisms underlying resistance. As plant breeding evacuates the universities, we can expect this to change to a more-focused, less-explained, but probably more cost-effective progress in disease resistance breeding.[118]*

Unfortunately, a fully commercialized plant-breeding program could seriously limit the possibilities for using resistant types. Genes incorporated into one proprietary plant variety may be unavailable to other breeders trying to improve resistance in another, possibly competing, variety. It may be well-nigh impossible to develop varieties specifically for use in integrated pest management programs—such as the short-season cotton bred in the early 1970s, which were central to insect management—since seed companies may have no particular incentive to breed such use-specific types. In the same vein, breeding multilines or other means of increasing the durability of resistance may be more difficult for business to achieve since more effort and resources would have to be committed to developing parallel lines without any guarantee of proportionately larger markets for the products. If, in addition, regional heterogeneity of crops or genetic types is needed to stabilize plant resistance, seed companies would be unlikely to support a strategy that could shrink markets.

> The only way to avoid dead ends is to maintain a vigorous public program of plant and animal breeding to complement or tie in with private research.

The only way to avoid these dead ends is to maintain a vigorous public program of plant and animal breeding to complement or tie in with private research. Publicly supported breeding should mainly seek to elucidate basic mechanisms of resistance to pests, develop comprehensive pest-management strategies incorporating host resistance, and meet the needs of those who grow fruit, vegetables, and other smaller-market commodities that may not attract enough private research capital. The findings of public programs belong to all scientists everywhere and help deepen our collective understanding of how pests and hosts interact and how to manage these interactions. Commercial breeders may bring what is already known to the market faster, appearing cost-effective in the short term. But public support of basic and applied research is still needed to find and explore what is not yet known. Without this public commitment, the private sector's apparent efficiency could be short-lived and our options for protecting crops and animals severely limited.

## Agroecology, Polyculture, and Pest Management

Biological strategies of pest management are now merging in so-called integrated farming systems.[119] Some approaches are based on the "organic" farming philosophy as it has evolved in industrialized countries,[120] while others come out of more formal applications of ecological principles to agricultural systems ("agroecology"), especially in developing countries.[121] Underlying all is an emphasis on the benefits of mixed cropping, or polyculture.

Organic (or "regenerative"[122]) farmers "assemble systems which are structured to achieve long-term stability."[123] Their first order of business is to maintain the health of the soil through rotation, recycling, and some natural amendments. They reject synthetic fertilizers and pesticides, but most use the same modern equipment and plant varieties as other farmers and sell their produce in most of the same markets.[124] To manage pests, organic farmers establish a healthy stand to ward off pests and disease. For weed control, most still rely on tillage, though some now reduce tillage without the use of herbicides. Some also use microbial pesticides and beneficial insects, but most commercial organic farms are too big to use these insects on crops cost-effectively. Most organic farmers rotate and mix crops carefully to maintain a healthy diversity. Crop rotation, used primarily to maintain soil fertility and proper soil structure, also helps control weeds, rootworms, and nematodes. Similarly, some mixed-crop tactics—such as overseeding maize with clover— enhance fertility and can shade out weeds, thus eliminating the need for repeated tillage or herbicide applications. Another tactic these farmers favor is leaving some non-crop plants on field borders to give refuge or alternative food sources to beneficial insects such as parasites and predators. These areas are pesticide-free, and natural biological controls can keep pest populations manageably low.[125] By the same token, the strong emphasis on soil health, the absence of toxic chemicals, and the frequent use of composted plant material and manure together suppress soil-borne pathogens and other pests.

Organic farming has only recently received serious attention from scientists and agricultural administrators.[126] In the United States, Congress is now considering a proposal to establish organic research/demonstration sites—a first, though such sites have existed for years in Europe. And the U.S. National Research Council has recently convened a panel to assess the possible contribution of organic-farming practices to solving some of agriculture's major problems. Detailed analysis of the success of organic practices may spark insights on how to make conventional agriculture more energy-efficient, reduce soil loss, and improve pest management. Similar agroecological investigations in the developing world need to be conducted on a wider scale to determine what elements of traditional agriculture need to be conserved in the transition to the highly productive systems needed to feed growing populations.

Agroecologists study organic farming but more often focus on traditional agriculture in the developing world. Often, they assume that traditional agroecosystems represent an optimal design, selected over millenia for stability and sustainability rather than maximum productivity.[127] Agroecological researchers look particularly closely at the effects, efficacy, and ecological underpinnings of mixed cropping. Apparently, the value of increasing plant diversity to manage pests depends on specific plant combinations rather than the increase *per se* in the number of species in a plot.[128]

Agroecological research and the growing scientific interest in organic farming bespeak a larger trend within agricultural science—a whole-systems approach to analyzing agriculture[129]—led in part by researchers in integrated pest management.[130] While biocontrol and resistant varieties may play useful roles regardless of other changes in agricultural practices, realizing their potential is easier if these tactics are carefully fit into a coordinated program. Conversely, if they aren't, biological methods of pest management could well fail. They could fall apart if pesticides aimed at controlling other pests are applied improperly or at the wrong time. If a resistant crop variety causes pest populations to fall low enough to starve important predators and parasites, natural controls could lose force. Or new pest-resistant varieties might require more fertilizer, which could cause nutritional changes that make the plant more attractive to other pests. Microbial pesticides may not work well unless applied efficiently and timed for maximum effect. In short, each technique's special characteristics and its interactions with all other elements of the management program must be examined systematically—the essence of integrated pest management.

# IV. Putting the Pieces Together: Integrated Pest Management

## What Is IPM?

In the late 1960s and the 1970s, accumulating evidence of the environmental hazards, health risks, and rising costs associated with the indiscriminate use of pesticides sparked policies and programs designed to curb the more harmful pesticide-use practices. These programs now include regulatory actions to restrict or remove some chemicals, training and certification programs to improve application safety, and research and demonstration projects in integrated pest management (IPM). In one of the best definitions of the IPM approach, the Office of Technology Assessment captured its key aspects:

> . . . the optimization of pest control in an economically and ecologically sound manner, accomplished by the coordinated use of multiple tactics to assure stable crop production and to maintain pest damage below the economic injury level while minimizing hazards to humans, animals, plants, and the environment.[131]

**IPM differs fundamentally from other approaches to controlling pesticides in two important respects. Economic benefits to the user are at least as important as environmental benefits to the wider community, and its emphasis is on redefining—not just refining—pest control strategies.**

IPM differs fundamentally from other approaches to controlling pesticides in two important respects. Economic benefits to the user are at least as important as environmental benefits to the wider community, and its emphasis is on redefining—not just refining—pest-control strategies. In IPM, each system (crop production system, public-health program, multiple-use forest, etc.) and its

43

particular needs is analyzed before those aspects of the system pertaining to pest problems are designed.

How did the IPM approach come about? Cotton production provides examples in several parts of the world, where over-reliance on pesticides made growing the crop well-nigh impossible. In northeastern Mexico, pesticide-resistant tobacco budworms forced farmers to stop growing cotton altogether, while in nearby Texas, as well as in Peru and parts of Central America, they came to rethink pest control, embracing IPM as an economical alternative to sole dependence on pesticides. Similarly, hopes that such insect-borne diseases as malaria would be eliminated from the earth have been dashed by resistance and resurgence; today, public-health specialists must adopt management, not eradication, as a goal. Often, IPM has evolved only when a pest-control program faced collapse. Concerns over pesticides' environmental effects have also been a factor. But in case after case, the top-to-bottom re-evaluation of pest control that characterizes the IPM approach has been undertaken only as a last resort.

Some of the many definitions of integrated pest management invite misconceptions. First, IPM is *not* the absence of chemical pesticides. While some chemical uses can be eliminated by substituting alternative methods, IPM can also sometimes require temporary increases in the use of certain chemicals. Second, IPM is *not* necessarily a combination of tactics, though often it is. Even when pesticides are the primary control method available, IPM can be used to manage the chemicals and the pests better. Third, IPM is *not* a return to old methods. It may involve reviving some sound practices hastily abandoned in the chemical revolution, but most new IPM techniques spring from fundamentally new analyses—*not* the "bad old days."

Since the late 1950s, the ideas underlying IPM have taken shape among agricultural scientists, principally entomologists. The key concepts are that there is a threshold level of damage below which control is not economically practical, that natural and chemical methods of pest control can be integrated, and that (since pests occur in populations that interact with other populations in complex associations) pest control must be grounded in ecology.

Unlike conventional pest-control approaches, IPM is based on a sound scientific understanding

> **Hopes that such insect-borne diseases as malaria would be eliminated from the earth have been dashed by resistance and resurgence; today, public-health specialists must adopt management, not eradication, as a goal.**

of the natural resource system being managed—the host (crop, animal herd, or human community), the pests, their natural enemies (predators, parasites, diseases), competitors, alternative hosts, etc. The social and economic reasons for managing the resource (say, farming for profit, managing a forest for public recreation or timber, etc.), and the incentives and constraints imposed by social, economic, political, and regulatory rules and values are also taken into account.

**IPM emphasizes systems design rather than knee-jerk reactions to pests.**

IPM emphasizes systems design rather than knee-jerk reactions to pests. Where possible, ecosystems are restructured to minimize the likelihood of pest damage. Another hallmark of IPM is optimization. Whether made mathematically or intuitively, the decision to apply control measures and the choice among them are made to balance the numerous costs, risks, and benefits involved. IPM programs also seek to stabilize pest-control systems by reducing fluctuations in pest populations, improving the predictability of control measures' effects, and making the control program more reliable over time. Instead of trying to eradicate pests from the environment, IPM systems try to maintain them at harmless levels.

In the end, IPM is economical to the user and to society both. In practice, users who adopt IPM in their own best economic interests bring wide-area environmental benefits to all. These double dividends are the essence of true optimization in pest management.

## Examples of IPM

*Cotton*

The control of insects on cotton has followed a similar pattern in much of the world. Synthetic organic pesticides are introduced, an initial flush of success follows, resistance soon builds up in the pests, the amounts of insecticide applied each season then increase rapidly while previously harmless species turn into major pests, and, finally, yields plummet and costs skyrocket.[132] Responses to this syndrome have varied, but all those that work have in common an area-wide approach that requires close cooperation among farmers. Two examples drive home the point.

The key pests of cotton in south-central Texas have long been the boll weevil and the pink bollworm. But, by the early 1960s, excessive insecticide spraying had so disrupted the natural controls in the cotton fields that another species c bollworm and the tobacco budworm—which had previously been innocuous—had also become serious problems. Although additional chemicals temporarily kept these new pests at bay, the tobacco budworm eventually developed such hig resistance to available insecticides that farmers faced ruin. Since then, an IPM program has put the budworm in its place. The main lines of defense have been the use of short-season cotton which helps control the boll weevil, and the elimination of second or third cuttings, which keeps the pink bollworm population from buildir up and eliminates a potential reservoir for a whitefly that transmits a damaging virus to cottor Other mainstays of the program include:

1. Irrigating prior to planting in desert areas if water is available;

2. Planting crops during a designated planting period (and fining uncooperative planters) that allows for maximum emergence of bollworm moths from their overwintering phase at a time when there is no cotton fruit available to lay eggs on;

3. Monitoring fields for pest and natural enemy population levels;

4. In areas where high boll weevil populations are anticipated, applying insecticide at a specific growth stage of the crop, which reduces the weevil population significantly but doesn't much harm natural enemies;

5. Defoliating or dessicating the mature crop with chemicals so that all cotton bolls open together, thus expediting machine harvesting;

6. Using mechanical strippers for harvest, whic kills most bollworm larvae;

7. Selectively applying an organophosphate insecticide during harvest to catch the adult boll weevil populations before they emigrate from the fields to overwintering sites in woods and field margins; and

8. Shredding stalks and plowing under the crop remnants immediately following harvest, to deny insects food and harbor over the winter

Together, these actions have yielded dramatic results. In two counties where harvested cotton acreage had been dropping from about 105,000 in 1970 to about 50,000 five years later, harvested acres increased to 236,500 once the program was in full swing. The area netted benefits estimated at $29 million from increased yields and decreased production costs.[133]

By the mid–1950s, the production of cotton in Peru's Cañete Valley had reached an all-time low, more and more insecticides were being applied, sometimes every two days, and the average plant still contained some 18-20 bollworm or budworm larvae. (The exploding population of another cotton-loving species was so numerous that the adults were depositing eggs on eucalyptus, doors of buildings, and automobile windshields!) The growers, at entomologists' urging, cooperated in a program that saw the crop recover within one to two years. This program had five elements:

1. Synthetic organic pesticides were banned with a return to older, more selective insecticides;

2. The practice of ratooning (bringing in a second harvest) was eliminated;

3. Uniform planting dates were established, and uncooperative farmers fined;

4. Insect parasites and predators were collected, reared and released; and

5. Maize and wheat were interplanted with cotton to facilitate the buildup of natural enemy populations.[134]

Both of these examples demonstrate the importance of regionally coordinated action among cotton-growers—one way to manage whole pest populations instead of trying to suppress them field-by-field in the proverbial losing battle. These two cases also illustrate the use of a combination of technologies to combat pests.

*Rice*

In Orissa, India, the gall midge, the brown planthopper, and several species of stemborers were damaging the state's rice crop in the 1970s. IPM experts first established economic thresholds—pest-density levels below which spraying isn't necessary—for the other major pests and developed a forecast system based on rainfall patterns to predict gall midge infestation. In the first year of this program, insecticide applications were cut in half. To achieve this,

1. Early-maturing, short-duration varieties of rice were sown to avoid those times when key pests were most numerous;

2. Insect levels were monitored to determine when pests exceeded threshold levels;

3. Varieties resistant to the gall midge or the stem borer were selected;

4. Crop residues were plowed under soon after harvest, preventing carryover of pest populations; and

5. No insecticides were applied when parasites and predators of the two key pests were most abundant.[135]

In the province of Kwangtung, in southern China, the Tahsia commune used an elaborate nine-point IPM program to reduce pesticide costs from $1.50 (U.S.) to $0.52 per acre:

1. Insect forecasting was added to the nationwide forecasting network;

2. Ducks were herded through the paddy fields to control weeds, leafhoppers, planthoppers, and leafrollers;

3. Rice fields were flooded to drown the larvae of the rice paddy borer;

4. Flying insects were trapped in large numbers with blacklight lures;

5. Disease-resistant varieties were planted;

6. Parasites of leafroller eggs were reared and released;

7. Frogs were conserved to prey on insects;

8. Fields were monitored and economic thresholds used to determine the need for insecticide spraying; and

9. *Bacillus thuringiensis* (B.t.) was applied to reduce the need for chemical insecticides.[136]

The Chinese approach to IPM requires great human labor to herd ducks, rear and release insect parasites, maintain blacklight traps, and the like—all possible where labor is plentiful. It also requires a high degree of social organization and coordination. The two-decades-old Chinese system of insect and plant-disease forecasting, for instance, requires monitoring insects' population density and dispersal, weather conditions, natural enemies, and other information, gathering the data at hundreds of stations around the country, and reporting regularly to provincial forecasting centers that prepare 10-day forecasts and project annual trends for the agricultural production units. And a national commitment to using biological control has meant that this option is seriously considered at every level.[137] Whether all of the well-coordinated elements of China's IPM program will continue amid the changing face of Chinese agriculture remains to be seen. So too, the applicability of the Chinese model to other developing countries is far more than a technical question. Nonetheless, the Chinese experience illustrates how a program can be comprehensive and sophisticated without necessarily relying on state-of-the-art equipment or college-educated farmers.

**The Chinese experience illustrates how a program can be comprehensive and sophisticated without necessarily relying on state-of-the-art equipment or college-educated farmers.**

*Maize*

Controlling corn rootworms is the cornerstone of Illinois' IPM program. When rootworms began growing resistant to the principal insecticides heptachlor and chlordane in the 1960s and unacceptable environmental and health risks cropped up, these chemicals were banned. Today, monitoring, crop rotation, and selective insecticide applications are used instead. Fields are scouted in August for rootworm densities. Where counts are high, soybeans or other crops that will not support rootworms are planted the following year. Where fields are left in corn, a soil insecticide is applied only if each plant hosts more than one rootworm on average. As a result, relatively less pesticide is needed, pesticide resistance hasn't built up in the rootworm populations, soil nitrogen is conserved, and some weeds are kept at bay without resort to herbicides. Even farmers who do not rotate their crops derive some benefit since the area's overall rootworm population declines.[138]

In Egypt, the cotton leafworm, the corn aphid, and three species of corn borers plagued the crop in the 1960s. Even three, sometimes four, insecticide applications per season didn't significantly curb losses. The IPM program introduced there in the mid-1970s centered on fixing sowing dates to escape most borer damage and setting economic thresholds to determine insecticide needs. By the late 1970s, this program had reduced insect-caused losses to about 3 percent, compared to over 10 percent in 1964. The area that had to be treated with chemicals to fight borers decreased from 692,000 acres to an average of 22,000. In addition, problems with leafworms, aphids, and mites declined substantially.[139]

*Fruit*

The key insect and disease pests of apples cannot be controlled effectively with available biological controls, mainly because some of these pests attack the fruit itself and farmers who expect to sell their produce in the fresh-fruit market can tolerate almost no damage. As a result, chemical pesticides remain the principal means of control, and economic thresholds have little meaning as applied to most pests. Yet, effective management strategies have been designed. In Michigan and elsewhere, these strategies have five components:

1. Weather and other environmental conditions that trigger outbreaks of apple scab and insect pests are monitored;

2. Key events in orchards (such as egg hatch and adult male insects' emergence from dormancy) are carefully watched;

3. The optimum times for insecticide applications are predicted using mathematical models;

4. The natural enemies of spider mites are conserved by using chemicals that are less harmful to the predators than the plant-feeders; and

5. Mite populations are monitored to decide when economic thresholds have been exceeded and selective pesticides are needed.[140]

Michigan's program pioneered the use of an extensive system of agricultural weather stations, computer terminals, scouting services, simulation models, and various communications capabilities to advise an area's apple-growers of prevailing pest conditions.[141] With this high-risk, high-value crop, optimizing pest management is difficult. But the regional approach has made the area's growers less apt to rely on regularly scheduled pesticide applications, and a new experimental program has shown that growers can further reduce spraying costs by deciding at the outset to sell their crop to food processors rather than fresh-fruit dealers.[142]

In California's San Joaquin Valley, a program for managing pests that infest citrus incorporates pest-population monitoring, damage assessment, reliance on predators and parasites, and the use of selective pesticides. The program, which has evolved since the mid-1960s as researchers learn more about the citrus ecosystem, has three cornerstones:

1. Key pests are monitored using sex-attractant traps, light traps, visual counts, and other methods;

2. Selective insecticides are applied when thresholds are exceeded; and

3. Natural enemies of mites and scale insects are protected.[143]

Citrus pests in Peru have been reduced by various methods designed to take full advantage of natural enemies. Mites have been effectively controlled with a combination of mineral oil and the botanical insecticide rotenone. The whitefly population is reduced initially by high-pressure water sprays, after which parasites do the job.[144]

Because so many pests attack fruit and so few consumers can abide even occasional blemishes, implementing IPM is harder for fruit-growers than for, say, cotton farmers. Still, pest management has made strides, and even when almost no pest damage can be tolerated, effective strategies can lessen dependence on the routine use of pesticides. By monitoring environmental conditions that favor pests' development and then precisely timing pesticide applications, IPM practitioners can reduce the amount of chemicals needed and preserve natural enemies. The handmaiden of success here is the predictive model, which helps growers time control measures and tells advisors which pests to monitor and when.

All the thumbnail case studies mentioned so far are agricultural and most concern mainly insects, principally because these programs are well documented. But the principles of IPM work equally well for managing insects and diseases in forestry, insects and weeds on rangeland and in

**The principles of IPM work equally well for managing insects and diseases in forestry, insects and weeds on rangeland and in pastures, insects on street trees, insects and plant diseases in suburban yards and urban parks, pests of livestock, rodents in cities, and insects that attack people.**

pastures, insects on street trees, insects and plant diseases in suburban yards and urban parks, pests of livestock, rodents in cities, and insects that attack people. In all of these cases, the keys are knowing as much about the pest and its environment as possible and using this knowledge to determine which natural factors govern the pest population's growth and decline and when to intervene in the pest's life cycle to prevent or reduce damage.

# V. Building Better Systems: Where Do We Go From Here?

## Systems and Components

Before the advent of synthetic organic pesticides, most pest-management problems were primarily *efficacy* problems—getting the products to work well. The chemical revolution after World War II "solved" these problems, but eventually led to *system* failures. Thus, for example, the buildup of resistance has not been a simple function of the chemical used or the physiology of the pest but, rather, a population phenomenon: resistance builds because the genetic, physiological, and ecological characteristics of populations interact in response to the effects of the chemical. Pest resurgence, secondary pest outbreaks, herbicide-induced succession, and the microbial breakdown of soil-applied pesticides are additional examples of unforeseen ecosystem responses to chemical assaults.

Human health effects and safety are systemic problems as well. It is not enough to carry out toxicity tests to decide which pesticides can be used and how. Pesticide application and formulation must be both efficacious and safe, which means that pesticide assessments are balancing acts. Most kinds of pesticide formulation have been tested only for their efficacy, not for their relative safety.[145] Nor have many pesticide products been properly tested for dislodgeable residues under the varying climatic conditions in which farmworkers are exposed.[146]

Most application methods are designed for convenience, economy, and efficacy, not to

> It is not enough to carry out toxicity tests to decide which pesticides can be used and how. Pesticide application and formulation must be both efficacious and safe, which means that pesticide assessments are balancing acts.

53

LORETTE WILMOT LIBRARY
NAZARETH COLLEGE

minimize the exposure to workers during mixing, loading, application, and cleaning. Yet, vast improvements are possible for increasing the efficiency of deposition on the target, possibly enhancing safety as well. Then too, since pesticide risk is a function both of toxicity and exposure, the dosage and frequency of application may significantly affect the human risk associated with pesticide use. Clearly, issues of health and safety cannot be neatly separated from those of ecosystem effects and responses in pest management.

Pests and the means to control them are not separate, independent entities, like pegs on a pegboard. Indeed, these "pegs" are connected to each other by overlapping webs of "string" of various strengths, thicknesses, and lengths. Rarely can a single peg be easily removed or replaced. As one ecologist put it, "you can't do one thing." Of course, some pest populations, crops, and pesticides, etc., can be identified, studied, and even modified as if they were distinct—the mainstay of most pest-control efforts since the chemical revolution began. Eventually, however, new or modified components must be fitted back into the system and it is then that the linkages—the webs in the pegboard—become important. If the scientist and the policy-maker don't consider these linkages, the farmer may have to find them out the hard way.

*Introduce a new pesticide. Later, face resistance or a new pest as beneficial species succumb to the pesticide. Introduce a biocontrol agent. Discover that a pesticide used to control another pest destroys the "good bug." Develop an improved plant variety only to find that a biotype of a disease can overcome that variety's resistance. Substitute one chemical for another, perhaps discovering the new chemical in groundwater because nobody knew what the chemical would do in different soil types.* In the evolution of IPM as an approach to solving problems like these, two types of research have emerged as equally important.[147] *Component research* covers the separate entities of pests, crops, and pest-control methods as before, though within the context of a system. *Systems research* is focussed on the behavior of the components and their linkages within the managed system. The goal is finding the most appropriate way to incorporate components into the system with the best results and the least harm. Most important, perhaps, systems research

**Systems research should reduce the "surprise" systemic failures that have been the hallmark of pest control in the chemical age.**

should reduce the "surprise" systemic failures that have been the hallmark of pest control in the chemical age. And systems thinking cannot stop with the researcher—practitioners and those who advise and regulate them must know that they are dealing with whole systems, not just one part.

As new pest-control methods are considered and developed, two sets of criteria should be used to evaluate them. As components, their efficacy and other potential effects can be assessed. As parts of systems, they raise other questions: How safe is a new pesticide to beneficial species? What is the pest's potential to develop resistance to a new chemical, or to overcome resistance in a new plant variety? What other species does a biocontrol agent affect? Will a resistant plant variety be more susceptible to another pest, thus requiring more pesticide? Will reducing one pest species help increase another? What will the net costs and benefits of introducing a given method be? Will variability decrease or increase? Will the technique require additional expertise on the user's part? Who will train or advise the users? Can users handle the method as intended? Will they use it? How risky will the method be to users in real-life, working conditions?

Since systems issues are difficult to tackle and easy to avoid, some of the gains made in the 1970s toward IPM were lost when the "new generation" of pesticides came along. Then too, many of the ecological and social concerns voiced in the 1960s and 1970s about agricultural trends have faded in the rush to join the coming biotechnology revolution.[148] The prospect of a new round of technological "breakthroughs" in pesticide development and plant breeding seems to be taking policy-makers and technology users alike back to a "magic bullet" philosophy—the belief that you *can* after all do just one thing. This return to individual technologies as godsends also stems from the private-product orientation in current research policy. The trend is especially evident in biotechnology, but it also shows up in the lack of interest in U.S. public research institutions for developing improved pesticide-application methods, classical biological controls, and plant breeding for pest resistance. With the "privatization" of research products, nobody is carefully assessing the system-level effects of new technologies and following up by monitoring and re-designing components.

55

Yet, the attraction of the product approach is obvious. Manufacturers of commercial products provide an army of advisors—salespeople and technical-support staff—at no apparent cost to the taxpayer. In contrast, IPM requires an extensive infrastructure—much of it publicly supported—to be effective. Moreover, there are no economies of scale in providing IPM advisory services and field-level monitoring, and the long-term rewards of using public funds to teach sound ecological principles to the next generation of farmers and pest-management consultants cannot compare with the lucrative payoffs that the biotechnology boom promises to industry-university liaisons.

Biotechnology, new chemistry, and other innovations have much to offer pest management. Selected and used carefully, they could stabilize rather than destabilize control, reduce rather than increase costs, and enhance rather than degrade the environment. Their promise is more sustainable, ecologically sound pest-management systems, not just another "patch-up" job.[149]

Of course, few of the technological choices policy-makers must make to accomplish these goals are value-neutral, and few should be left solely to market forces—an invitation to a new round of systemic "surprises." Rather than waiting to be surprised, policy-makers could begin programs of evaluation, monitoring, advising, and system-level research now so that the most-needed technologies will get developed and used appropriately. More than any other factor, the further evolution of an effective IPM infrastructure is essential to this objective.

## An IPM Support System

As the last chapter shows, IPM is not a technology but a design and decision-making process for structuring ecosystems to minimize pest damage and coping better with unavoidable pest problems. While IPM changes as information, conditions, and technologies change, the criteria for judging its effectiveness—productivity, stability, sustainability, and equity—do not.

In practical terms, the key to an effective IPM approach is a solid support structure for design, monitoring, and advice. Such a comprehensive support system has several functions: research, reporting, delivery, and education.[150] Its objective should be not only to develop new technologies but also to pinpoint the needs for new methods and assess their effects.

Assessing system-level effects must become a high public priority in pest management. When a new technology is proposed for use in a management system, its potential for disrupting as well as improving management must be carefully considered. Currently, pesticide regulators pay little attention to how chemicals affect ecosystems[151] or to how pesticides affect such beneficial species as the natural enemies of pests.[152] Similarly, predicting resistance to pesticides is crude at best,[153] and means for anticipating the durability of host-plant resistance to pests are sorely needed.

As for IPM research, technology development needs both a push from the present and a pull from the future if pest management is to be stabilized. On the one hand, farmers and other technology users and their immediate advisors should together set the short-term research agenda. On the other, an appropriate long-term research course will emerge only if policy analysts and others concerned with global and regional trends in resources, population, and environment get involved.

Translating broad directions into practicable pest-management strategies in the face of great uncertainty requires considerable creativity. It also requires public support since social goals and goods are at stake. Yet business has its role too. Private R&D is needed to develop the specific products that will fit into an overall design. Deciding the appropriateness of particular technologies is a function of the advisor or the farmer who must bring techniques together into a working "package." In many countries, this integrating will be done by private consulting companies who advise farmers how best to manage their crops and pests. In others, government scientists will have to carry out this critical step.

The reporting function in a comprehensive IPM support system, its intelligence phase, tells us where we stand and where we are going. Area-wide data on pest populations, pesticide resistance levels, pesticide usage, performance of pest-resistant plant varieties, and crop losses,[154] effects on "nontarget" organisms, occupational health, and pesticide residues in food and the environment are needed to analyze research needs, assess the effects of innovations, and improve the efficiency and effectiveness of IPM delivery. They can allow more precise timing of field-level monitoring for pest-management decisions and help check the wide-area effects of field-level decisions. Good information-gathering should also help keep the "external" costs of pest management down, permitting societies to respond quickly to problems generated by individual pest-control actions. Reporting need not be centralized, but it must be well coordinated or cross-cutting analyses and comparisons become impossible. No data base or data bank can accommodate all possible combinations and permutations of information, but agreement on terms, units of measurement, recording protocols, and quality assurance can make even a diffuse information system usable.

At the heart of IPM support is the delivery system. More than simply education and training, delivery implies a full spectrum of services including monitoring, short-term forecasting, and advice. The delivery system may be greatly enhanced with computers and telecommunications, but these are not essential to begin an effective IPM program. It may even be necessary to train pest managers to use advice to greatest advantage. Field-level monitoring of pests, pesticides, and crop losses also supplies a significant portion of the regional-level data bases managed in the reporting function. Users' observations are essential too.[155] The professionals in IPM delivery should help link users to researchers and policy-makers, helping to communicate users' needs, practices, and motivations. Delivery can be either public or private. Indeed, in many areas, independence from government or corporate control may be essential to a program's quality and success. Local farmers' organizations, especially pest-management cooperatives,[156] may work best in this capacity where the market for IPM services cannot support individual entrepreneurs.

No IPM support system can remain viable for long without programs for educating researchers, advisors, and others who provide sustained support. Just as a human health-care system depends on a steady stream of well-trained physicians, nurses, and technicians, a "plant-health system"[157] requires a cadre of professionals to advise users, design and conduct monitoring

**Just as a human health-care system depends on a steady stream of well-trained physicians, nurses, and technicians, a "plant-health system" requires a cadre of professionals to advise users, design and conduct monitoring programs, and develop new technologies.**

programs, and develop new technologies. With IPM in its infancy, that cadre is still small. Its growth can be accelerated by public policies designed to institutionalize IPM in agriculture and elsewhere. For instance, a phased plan to require prescriptions for using certain high-risk pesticides would help generate a market for advisory services, providing an incentive for students to prepare for careers in IPM[158] (and helping assure that pesticide use does not exceed need[159]). Similarly, establishing national and regional monitoring programs would both build the crucial data bases needed to improve IPM systems and generate jobs for technically trained personnel. Governments can also require public agencies to implement IPM in their operations wherever appropriate[160] and to fund needed research and monitoring. The knowledge and information gained in these efforts could help upgrade nongovernmental IPM programs as well, opening new opportunities for IPM professionals.

## Initiatives to Build an IPM Support System

Where market forces alone cannot bring effective and environmentally sound pest-management technologies into use, government's role is legitimate and important. In the case of inoculative biological control, for instance, government should support the research, importation, and establishment of such beneficial species and prohibit private actions (such as the use of certain pesticides) that disrupt those species. If private companies do not find breeding for pest resistance in certain crops economically attractive, new varieties could be developed with public funds. Standards for varietal performance could be set by state or national agriculture departments that require acceptable levels of resistance to pests, and plant varieties developed by private or public laboratories should meet the same standards. Governments could also do more to direct biotechnology development related to pest management, either by funding or subsidizing promising research areas on control methods or by discouraging or preventing the development of techniques that could destabilize or degrade IPM systems.

How should this new IPM infrastructure be financed? In an earlier World Resources Institute study,[161] a sales tax on pesticides was proposed as a means of paying for resistance research and monitoring. A public trust fund based on such a nominal tax could defray most of the near-term costs of developing IPM. At the same time, commodity associations or farmer cooperatives could fund crop-specific research, monitoring, and similar activities.

In developing countries, where financial resources are chronically short, aid from multilateral and national agencies should be used to create and support an IPM support system as part of an overall effort to improve and stabilize crop production. Then too, short-term aid in battles against particular pests should be tied to longer-term solutions, including the buildup of an effective infrastructure. Where national institutions are chronically weak in pest-management and related disciplines, regional institutes and international research centers could be used for area-wide pest-management research, reporting, delivery, and training, with funds from traditional international sources.

As for the role of new technologies in pest management, their success will depend mainly on how they are used. Without IPM support systems like the one described here, the potential of new pest-management methods could well be undercut. Beneficial technologies may be misused or ignored if no one trains users. Perhaps worse, great ideas may go undeveloped if a lack of supporting institutions makes markets look weak or implementation difficult. For many methods, improper application may spoil the opportunity forever, as has happened with pesticide resistance. For others, failure might mean the waste of scarce research and development resources, with concomitant delay in progress against persistent agricultural or public-health pests.

A long-term strategy for developing and implementing improved pest-management systems must rest on structured, rational decision-making. Without it, pest management can't evolve in a stable fashion and incorporate new technologies as they prove useful. Products—whether biological, chemical, biotechnological, or mechanical—can and will be developed to meet identified needs, but who decides how those products will be used? And how will the decisions be made? Surely, governments cannot leave this matter entirely to the marketplace. Just as the airline industry

**Just as the airline industry depends on publicly-supported weather monitoring and forecasting, air traffic control, and safety research and regulation, so too pest management and its associated industries require the research, educational, monitoring, and advisory capacities that can only reside in the public sector.**

depends on publicly-supported weather monitoring and forecasting, air traffic control, and safety research and regulation, so too does pest management and its associated industries require the research, educational, monitoring, and advisory capacities that can only reside in the public sector. And just as human-health and animal-health systems require trained professionals, publicly funded research, and some government regulation, a plant-health system needs staff, expertise, and knowledge equal to the task of effective national and international management. Without these support mechanisms, progress will be sporadic and incremental at best.

This blueprint for implementing IPM should give policy-makers guidance and a durable approach to decision-making. But specific initiatives in research, extension, regulation, and international assistance are also needed.

## Research

*Expand systems-oriented interdisciplinary research on pest management.*

A team approach to pest-management research has proven itself as the most effective method of organizing IPM research. In the United States, the 16-university Consortium for Integrated Pest Management has made impressive gains in developing IPM systems for alfalfa, apples, cotton, and soybeans. The U.S. Department of Agriculture's program in Regional Projects for IPM is also organized along interdisciplinary lines, and this (albeit slow-starting) program deserves support. Indeed, other national research programs could be modeled after the USDA program. Many of the International Agricultural Research Centers, commodity-oriented to begin with, are organized in a similar fashion. Currently, these centers focus primarily on plant breeding for increased production, and pest management's potential in maintaining net production gains is sometimes overlooked. But, a team structure—the basis for a solid systems approach to pest management—has already been established in the international centers, so IPM can move forward there with a relatively simple change of emphasis.

*Upgrade and expand biological-control research in international and national research centers.*

Especially in developing countries, where pest management sometimes depends less on pesticides

61

than in the more industrialized countries, the potential for applying biological controls to pest problems is great. Opportunities are especially promising where plant breeding has successfully provided full or partial resistance to certain pests in major cropping systems. Biological-control research should emphasize all three strategies—inoculative release, inundative release, and conservation of existing natural enemies. Priority should be given to inoculative release since it is essentially permanent and offers the highest payoff. Pest-management programs using inundative release can create opportunities for small businesses and farmer cooperatives in developing countries to provide products and services locally with relatively small investments of capital. Research to develop economical means of mass-rearing organisms for inundative release is needed as well as new methods for properly timing and dispersing the biocontrol agents. Technical assistance may be needed to help prevent misapplication and overuse, which can precipitate system failures. International research centers can do much to further such research. For instance, the International Centre of Insect Physiology and Ecology in Kenya devotes a major portion of its research to discovering and developing biocontrol agents. But national and regional work is also needed to assure that biocontrol candidates fit well into local ecosystems.

*Contract biotechnology companies to develop and produce microbial pesticides for small markets.*
Like the producers of conventional pesticides, biotechnology companies can be expected to develop microbial products principally for the major markets—cotton, corn, soybeans, and a few other crops—and principally for use in developed countries. While Third World markets may be of secondary interest to these firms, the potential for improving pest control in both staple and specialty crops may greatly benefit developing countries. One solution to this impasse—short of establishing large numbers of potentially duplicative national and international research efforts in biotechnology-is to identify promising organisms for development (through existing agricultural research centers in developing countries) and then to contract biotechnology companies to develop products and production methods aimed at producing and distributing the microbes in countries where they

can be used. This two-step technology transfer could be funded through international assistance agreements, and some of the costs might be offset if the projects were conducted as joint ventures between the agricultural research centers and the companies involved.

*Develop more and better methods for assessing pesticides' effects on biological control agents.*

Pesticide registration requirements must eventually be expanded to include data on the risks to pests' natural enemies. Additionally, whether biocontrol agents being screened for possible field use can survive in pesticide-treated environments needs to be evaluated. For both purposes, methods are needed for effectively determining the safety of how pesticide uses affect beneficial species. This work should be directed and coordinated by both regulatory and agricultural agencies.

*Expand and coordinate research on assessing exposure to pesticides.*

Developing effective means of reducing pesticide exposure cannot await a fuller understanding of the wide-ranging factors that affect exposure. But further research on exposure is needed, and methods for assessing exposure need to be improved and standardized. While the work on exposure should be conducted by pesticide companies and by university-based agricultural scientists, agencies concerned with occupational safety and health should take overall responsibility for setting research guidelines and coordinating research objectives.[162]

*Give competitive research grants for improving safety in pesticide application.*

Because the market for improved pesticide-application equipment is limited, few companies invest in innovative research. Publicly-funded engineering research could significantly improve application technology. A well-funded national competitive grants program could go a long way toward bettering application methods. Support for such a program could come from a combination of general funds, the previously mentioned tax on pesticides, and a modest tax on current application equipment. Innovative equipment that demonstrably reduces drift or otherwise lessens exposure could be exempted from this equipment tax.

# Extension

*Re-examine the structure and functions of national extension programs.*

Since every nation's and state's extension service has its own special character, proposing generic changes in extension practices makes little sense. However, the experiences of several state extension services in the United States in delivering IPM training and information, and the success in parts of Asia of the Training and Visit system,[163] underscore the need to carefully examine extension services' role in IPM. Traditionally, extension has been synonymous with education, and the extension agent's role with that of teacher and trainer. But IPM requires more: an active, responsive body that monitors pest populations and provides timely advice to users on the best alternatives for optimal pest management. In some areas, the extension service may be the only institution capable of performing these duties. In any case, every country should evaluate its service with the availability of alternative advisory services, the sophistication of users, and the skills and duties of extension agents and specialists in mind.

# Regulation and Incentives

*Require the use of closed systems for transferring highly toxic pesticides.*

Experience in California has demonstrated that the use of closed systems can reduce the incidence of accidental poisoning among mixers and loaders, the group most at risk from pesticide exposure. Partly because of California's requirements, new technology is now available, so there is no reason not to implement similar restrictions wherever such chemicals are used.

*Set standards and timetables for reducing drift in pesticide application.*

Besides sponsoring research to improve application efficiency, governments can give pesticide users and equipment manufacturers a clear deadline for meeting specific standards for drift reduction. Given the diversity of equipment and the potential for modifying machinery and methods to reduce drift, the choice of technologies need not be specified in such standards. But a target level of allowable drift will help get modified nozzles, drift shields, electrostatic sprayers, and other drift controls into the field.

*Require data on pesticides' effects on biological control agents as part of pesticide registration.*

While additional research on safety to beneficial species is needed, enough information now exists to establish initial data requirements with current methods. In the United States, the U.S. Environmental Protection Agency (EPA) has so far declined to issue such guidelines—an unfortunate lapse, considering how many countries watch and follow the United States. EPA has two alternatives: it could specify a few representative species for testing or, preferably, it could list the key beneficial species in major crop systems and require registrants to test their products for safety to these species under field conditions as a requirement for registering them for use on those crops. This second alternative could be accomplished as part of expanded studies under the ''experimental use permit'' or its equivalent.

*Restrict high-risk pesticides to use within approved IPM programs.*

If a pesticide's safety or environmental effects are in doubt, the compound's use might be continued only in programs that minimize risk and maximize benefits. If IPM programs that reduce exposure or otherwise provide a favorable benefit/risk ratio can be identified, then pesticides that might otherwise have to be shelved could continue to be used.

*Discontinue subsidization of pesticide prices.*

A survey of eight developing countries conducted by the World Resources Institute found that pesticides were subsidized by governments at an average of 44 percent of the real price.[164] Such subsidies encourage overuse of pesticides and give chemicals an unrealistic economic advantage over alternative methods. Subsidies also undermine the central tenet of integrated pest management— that pests should be controlled only if they exceed an economic threshold. If the costs of control (pesticide prices) are artificially low, then the incentive to spray will not be constrained by the need to look at the real economic benefits.

Governments that subsidize pesticide prices should consider other ways to encourage increased production. The sizable sums spent on

65

pesticide subsidies could be applied to enhancing IPM research, extension, and monitoring with far greater long-term benefits. If incentives in pest management are needed at all, it is in getting farmers to use sound, balanced IPM advice. Providing such advisory services, with a definite timetable for phasing out subsidies, could help bring about more rapid adoption of IPM.

## International Assistance

*Require assessments of pest-management needs and pesticide-application methods as part of the evaluation of proposed agricultural development projects.*

Multilateral and bilateral aid agencies can apply considerable leverage when their recipients decide what types of technologies to use in development projects. Donors and lenders could require that both the efficacy and safety of pest-management activities be monitored in the projects they support. More specifically, assistance organizations should make aid contingent upon a thorough analysis of the pest problems that face a particular project and of the project's capabilities for handling them. Such an assessment should not only list any pesticides proposed for use, but also describe who will decide how, why, when, where, how much, and how often chemicals will be applied. Additionally, the pesticide-application equipment available for the project should be surveyed and the expected safety levels that can be achieved with that equipment indicated.

*Set standards for the pesticide-application equipment used in agricultural development projects and provide funds to upgrade equipment to meet those standards.*

Determining pesticide-exposure levels and relating them to different types of equipment is both tedious and tricky, and few developing-country governments have the money and trained manpower needed to set standards for application equipment. But that shouldn't prevent assistance organizations from working with environmental agencies in industrialized countries to develop such standards and use them as a basis for funding equipment to be used in projects. Where equipment doesn't meet safety standards, the purchase of new application devices might be made a condition of project support.

*Provide funds to support monitoring and advisory capabilities for IPM.*
Creating and maintaining a network of trained IPM personnel in developing countries will take both time and money. But over the long term, support for monitoring, advisory services, and training will help improve safety, reduce operating costs for farmers, and improve net yields. If funding for monitoring and advice is inadequate, IPM research funds will be wasted because the results cannot be readily implemented. Failure to provide these services will also lead to overuse of pesticides and to increasing health and environmental risks in countries that can ill afford them.

*Provide funds for establishing and maintaining national-level research and quarantine facilities for biological control.*
Inoculative release is probably the safest, most cost-effective pest-management strategy available. Virtually all that limits its success are effort and the availability of proper facilities for research and screening. Establishing a broader international network of well-staffed research and quarantine facilities for biological control will help developing and industrialized countries alike since many pests in the North come initially from the tropics. Reciprocal arrangements for conducting surveys and research and for developing production methods could greatly enhance the viability of biocontrol everywhere.

***

The pesticide dilemma is that every attempt to take one control action causes numerous unintended effects. Since the 1940s, this dilemma has often led to a treadmill effect: the more we use certain chemicals, the more of them we have to use, and the worse the problem becomes. Human health has been put at risk, untoward ecological consequences have occurred, and costs have often gone out of control. Yet, used carefully and judiciously, pesticides can improve crop production and enhance public health. Technologies now being developed can reduce exposure to pesticides by better controlling their application, reducing rates and frequency of use, and eliminating unnecessary uses. The key to success is recognizing that pests and pest control take place in a larger system. That system must be understood and intelligently managed if ugly surprises are to be avoided and the full benefits of pest management realized.

**Dr. Michael J. Dover** is an Associate in WRI's Pesticide Project. Formerly, he served in the Office of Pesticide Programs and the Office of Toxic Substances at the Environmental Protection Agency.

**Notes**

1. Douglas G. Baugher, "Exposure to Pesticides During Application: A Critical Review of the State of the Art," consultant report to World Resources Institute, October 1984.

2. John E. Davies, "Health Effects of the Global Use of Pesticides," report prepared for World Resources Institute, 1985.

3. *Ibid.*

4. Council on Environmental Quality, *Environmental Quality* (President's Council on Environmental Quality, 1982).

5. Robert F. Wasserstrom and Richard Wiles, *Field Duty: U.S. Farmworkers and Pesticide Safety* (Washington, D.C.: World Resources Institute, 1985).

6. Jorge Manring, "Ecological Effects of Pesticides," consultant's report to World Resources Institute, 1985.

7. Michael Dover and Brian Croft, *Getting Tough: Public Policy and the Management of Pesticide Resistance* (Washington, D.C.: World Resources Institute, 1984); George P. Georghiou, "The Consequences of Evolutionary Adaptation of Pests to Pesticides," presented at Symposium on Pesticide Resistance Management, National Academy of Sciences, November 27–28, 1984; Norman Gratz, comments to the Symposium on Pesticide Resistance Management.

8. Mary Lou Flint and Robert van den Bosch, *Introduction to Integrated Pest Management* (New York: Plenum Press, 1981).

9. *Ibid.*

10. David Pimentel and Marcia Pimentel, "Dimensions of the World Food Problem and Losses to Pests," in David Pimentel, ed., *World Food, Pest Losses, and the Environment* (Boulder, Colo.: Westview Press, 1978), pp. 1–16.

11. Ray F. Smith and Donald J. Calvert, "Insect Pest Losses and the Dimensions of the World Food Problem," in Pimentel, *World Food, Pest Losses, and the Environment,* pp. 17–38.

12. J. Lawrence Apple, "Impact of Plant Disease on World Food Production," in Pimentel, *World Food, Pest Losses, and the Environment,* pp. 39–50.

13. David Pimentel, "World Food Crisis: Energy and Pests," *Bulletin of the Entomological Society of America* 22 (1976): 20–26.

14. Dover and Croft, *Getting Tough.*

15. *Ibid.*

16. Frank R. Hall, introduction to Agricultural Research Institute (ARI) Conference on Improving Agrochemical and Fertilizer Application Technology, Chevy Chase, MD, February 27–March 1, 1985.

17. Baugher, "Exposure to Pesticides During Application"; Wasserstrom and Wiles, *Field Duty.*

18. See, for example, Francis A. Gunther and Jane D. Gunther, eds., "Residues of Pesticides and Other Contaminants in the Total Environment," *Residue Reviews,* volume 75, 1980; Jack R. Plimmer, ed., *Pesticide Residues and Exposure,* ACS Symposium Series 182 (Washington, D.C.: American Chemical Society, 1982); Joseph V. Rodricks and Robert G. Tardiff, eds., *Assessment and Management of Chemical Risks,* ACS Symposium Series 239 (Washington, D.C.: American Chemical Society, 1984); Division of Pesticide Chemistry, *Program Abstracts,* 187th ACS National Meeting, St. Louis, MO, April 8–13, 1984 (Washington, D.C.: American Chemical Society, 1984).

19. G.S. Batchelor and K.C. Walker, "Health Hazards Involved in the Use of Parathion in Fruit Orchards in North Central Washington," *American Medical Association Archives of Industrial Hygiene* 10 (1954): 522.

20. W.F. Durham and H.R. Wolfe, "Measurement of the Exposure of Workers to Pesticides," *Bulletin of the World Health Organization* 26 (1962): 75.

21. Baugher, "Exposure to Pesticides During Application."

22. *Ibid.*

23. Wasserstrom and Wiles, *Field Duty.*

24. M.E. Whalon and B.A. Croft, "Apple IPM Implementation in North America," *Annual Review of Entomology* 29 (1984): 435–70.

25. U.S. Congress, Office of Technology Assessment, *Pest Management Strategies in Crop Protection* (Washington, D.C.: U.S. Government Printing Office, 1979).

26. Wasserstrom and Wiles, *Field Duty.*

27. Baugher, "Exposure to Pesticides During Application"; A.C. Waldron, "Minimizing the Pesticide Exposure Risk for the Mixer-Loader and Applicator," in *Program Abstracts,* 187th ACS National Meeting.

28. Molly J. Coye, "The Health Effects of Agricultural Production," unpublished manuscript, 1984.

29. J.E. Davies et al., "Protective Clothing Studies in the Field: An Alternative to Reentry," in Plimmer, ed., *Pesticide Residues and Exposure,* pp. 169–182.

30. Joan Laughlin and Carol Easley, "Methyl Parathion Residues in Contaminated Fabrics After Laundering," in *Program Abstracts,* 187th ACS National Meeting; J.O. DeJonge et al., "Protective Apparel Research," in *Program Abstracts,* 187th ACS National Meeting.

31. Davies et al., "Protective Clothing Studies in the Field."

32. DeJonge et al., "Protective Apparel Research."

33. Wasserstrom and Wiles, *Field Duty.*

34. ARI Conference on Improving Application Technology, various speakers; however, it is worth noting one of the conference objectives was to identify "[p]rinciples and practices in risk management associated with the development and utilization of new chemical application technology."

35. G.A. Matthews, *Pesticide Application Methods* (New York: Longman Group, 1979).

36. Baugher, "Exposure to Pesticides During Application."

37. *Ibid.*

38. John Maybank, "Transport, Drift Characterization, Diffusion and Exposure," presented at ARI Conference on Improving Application Technology.

39. Matthews, *Pesticide Application Methods*; Dale E. Gandrud, presentation at ARI Conference on Improving Application Technology; Dale E. Gandrud and John H. Skoglund, "Advances in Dry Pesticide Application Technology: Dry Application of Dry Flowable Formulations," presented at Symposium on Advances in Pesticide Application Technology, American Chemical Society, Philadelphia, PA, August 30, 1984.

40. Matthews, *Pesticide Application Methods*; Jean H. Dawson, presentation at ARI Conference on Improving Application Technology.

41. William Hairston, presentation at ARI Conference on Improving Application Technology.

42. Baugher, "Exposure to Pesticides During Application."

43. T.B. Hart, "The Hand-Held 'Electrodyn' Sprayer: Worker Hazard," ICI Plant Protection Division (Fernhurst, England, undated); R.A. Coffee, "Electrodynamic Crop Spraying," *Outlook on Agriculture* 10 (1981): 350–56.

44. D.L. Reicherd and T.L. Ladd, "Pesticide Injection and Transfer System for Field Sprayers," *Transactions of the American Society of Agricultural Engineers* 26 (1983): 683–86; T.L. Ladd and D.L. Reicherd, "Injection-Type Field Sprayer for Control of Insect Pests," *Journal of Economic Entomology* 76 (1983): 930–32; M. Schmidt, "The Direct Injection Technique for Preparing the Spray Mix—A Method for Reducing Safety and Hygiene Problems in Plant Protection?", *OEPP/EPPO Bulletin* 13 (1983): 513–20.

45. Anonymous, "Pesticide Packaging and Communication," *Infoletter* (Newsletter published by the International Plant Protection Center, Oregon State University), No. 58 (1983).

46. W.E. Yates, N.B. Akesson, and R.W. Brazelton, "Systems for the Safe Use of Pesticides," *Outlook on Agriculture* 10 (1981): 321–26.

47. *Ibid.*

48. Vic Thorpe and Nigel Dudley, "Pall of Poison: The 'Spray Drift' Problem" (Stowmarket, U.K.: The Soil Association, 1984).

49. David Pimentel, private communication.

50. Gandrud and Skoglund, "Advances in Dry Pesticide Application Technology."

51. Barry Rogers, presentation at ARI Conference on Improving Application Technology.

52. Graham A. Matthews, "International Developments/ Innovation," presented at ARI Conference on Improving Application Technology.

53. Coffee, "Electrodynamic Crop Spraying"; S. Edward Law and Harry A. Mills, "Electrostatic Application of Low-Volume Microbial Insecticide Spray on Broccoli Plants," *Journal of the American Society for Horticultural Science* 105 (1980): 774–77; Donald G. Manley, "Use of an Electrostatic Sprayer for Cotton Insect Control," *Journal of Economic Entomology* 75 (1982): 655–56; M.D. Lane et al., "Electrostatic Deposition Technology for Spraying of Vegetables," presented at 188th National Meeting of the American Chemical Society, Philadelphia, PA, August 26–31, 1984.

54. Richard P. Reynolds, "Report and Guidelines for the Packaging and Storage of Pesticides," Report to the Food and Agriculture Organization of the United Nations, January 1982.

55. Coffee, "Electrodynamic Crop Spraying."

56. Gandrud and Skoglund, "Advances in Dry Pesticide Application Technology."

57. Coffee, "Electrodynamic Crop Spraying."

58. Baugher, "Exposure to Pesticides During Application."

59. Hall, introduction to ARI Conference.

60. Loren E. Bode, presentation to ARI Conference on Improving Application Technology.

61. Baugher, "Exposure to Pesticides During Application."

62. M. John Bukovac, "Spray Technology—Shortcomings and Opportunities with Special Reference to Tree Fruits," presented at ARI Conference on Improving Application Technology.

63. Bode, presentation to ARI Conference.

64. Graham A. Matthews, comments at ARI Conference on Improving Application Technology.

65. Dover and Croft, *Getting Tough*.

66. Office of Technology Assessment, *Pest Management Strategies*; Dale R. Bottrell, *Integrated Pest Management* (Washington, D.C.: President's Council on Environmental Quality, 1979).

67. Office of Technology Assessment, *Pest Management Strategies*.

68. Thomas C. Edens and Dean L. Haynes, "Closed System Agriculture: Resource Constraints, Management Options, and Design Alternatives," *Annual Review of Phytopathology*, 20 (1982): 363-95.

69. Gordon R. Conway, "Agroecosystem Analysis" (London: Imperial College Centre for Environmental Technology, 1983).

70. J.C. van Lenteren, "The Potential of Entomophagous Parasites for Pest Control," *Agriculture, Ecosystems and Environment* 10 (1983): 143-58; F.W. Stehr, "Parasitoids and Predators in Pest Management," in R.L. Metcalf and W.H. Luckmann, eds., *Introduction to Insect Pest Management* (New York: John Wiley and Sons, 1982), pp. 135-73.

71. M.E. Brown and J.E. Berenger, "The Potential of Antagonists for Fungal Control," *Agriculture, Ecosystems and Environment* 10 (1983): 127-42.

72. Stehr, "Parasitoids and Predators in Pest Management."

73. Suzanne W.T. Batra, "Biological Control in Agroecosystems," *Science* 215 (1982): 134-39.

74. Van Lenteren, "The Potential of Entomophagous Parasites for Pest Control."

75. David Pimentel et al., "Pesticides: Environmental and Social Costs," in David Pimentel and John H. Perkins, eds., *Pest Control: Cultural, and Environmental Aspects* (Boulder, Colo.: Westview Press, 1980), pp. 99-158.

76. Batra, "Biological Control in Agroecosystems."

77. H.C. Coppel and J.W. Mertins, *Biological Insect Suppression* (Berlin: Springer Verlag, 1977).

78. Heikki Hokkanen and David Pimentel, "New Approach for Selecting Biological Control Agents," *Canadian Entomologist* 116 (1984): 1109-21.

79. R.P. Jacques, "The Potential of Pathogens for Pest Control," *Agriculture, Ecosystems and Environment* 10 (1983): 101–26; D.W. Roberts, ed., *Proceedings of Workshop on Insect Pest Management with Microbial Agents* (Ithaca, New York: Boyce Thompson Institute, 1980).

80. Jacques, "The Potential of Pathogens for Pest Control."

81. J.E. Henry and J.A. Onsager, "Large-Scale Test of Control of Grasshoppers on Rangeland with *Nosema locustae*," *Journal of Economic Entomology* 75 (1982): 31–35.

82. George E. Templeton and David O. TeBeest, "Biological Weed Control with Mycoherbicides," *Annual Review of Phytopathology* 17 (1979): 301–10.

83. Brown and Berenger, "The Potential of Antagonists for Fungal Control"; Kenneth F. Baker, "The Future of Biological and Cultural Control of Plant Disease," in Thor Kommedahl and Paul H. Williams, eds., *Challenging Problems in Plant Health* (St. Paul, Minn.: American Phytopathological Society, 1983), pp. 422–30.

84. Larry G. Bezark and Eric J. Rey, "Suppliers of Beneficial Organisms in North America" (Sacramento, Calif.: State of California Department of Food and Agriculture, 1982).

85. T. Boza Barducci, "Ecological Consequences of Pesticides Used for the Control of Cotton Insects in Cañete Valley, Peru," in M.T. Farvar and J.P. Milton, eds., *The Careless Technology: Ecology and International Development* (New York: Natural History Publications, 1972), pp. 423–38.

86. Qu Geping, "Biological Control of Pests," in Qu Geping and Woyen Lee, eds., *Managing the Environment in China* (Dublin: Tycooly International, 1984), pp. 87–94; Qu Geping, "Biological Control of Pests in China," *Mazingira*, 7 (1983): 24–31.

87. Van Lenteren, "The Potential of Entomophagous Parasites for Pest Control."

88. Katherine H. Reichelderfer, *Economic Feasibility of a Biological Control Technology* (Washington, D.C.: U.S. Department of Agriculture, 1979).

89. Batra, "Biological Control in Agroecosystems."

90. N.H. Starler and R.L. Ridgway, "Economic and Social Considerations for the Utilization of

Augmentation of Natural Enemies," in R.L. Ridgway and S.B. Vinson, eds., *Biological Control by Augmentation of Natural Enemies* (New York: Plenum Press, 1977), pp. 431-53.

91. Jacques, "The Potential of Pathogens for Pest Control."

92. Stehr, "Parasitoids and Predators in Pest Management"; Batra, "Biological Control in Agroecosystems."

93. Whalon and Croft, "Apple IPM Implementation in North America."

94. P.L. Adkisson et al., "Controlling Cotton's Insect Pests: A New System," *Science* 216 (1982): 19-22.

95. Carl B. Huffaker, ed., *New Technology of Pest Control* (New York: John Wiley and Sons, 1980).

96. U.S. Department of Agriculture Study Team on Organic Farming, *Report and Recommendations on Organic Farming* (Washington, D.C.: U.S. Government Printing Office, 1980).

97. David Andow, "The Extent of Monoculture and its Effects on Insect Pest Populations with Particular Reference to Wheat and Cotton," *Agriculture, Ecosystems and Environment* 9 (1983): 25-36.

98. Dover and Croft, *Getting Tough*.

99. Whalon and Croft, "Apple IPM Implementation in North America."

100. Dean L. Haynes, R. Lal Tummala and Thomas C. Ellis, "Ecosystem Management for Pest Control," *BioScience* 30 (1981): 690-96.

101. D.J. Greathead and J.K. Waage, *Opportunities for Biological Control of Agricultural Pests in Developing Countries* (Washington, D.C.: The World Bank, 1983).

102. Paul DeBach, quoted in van Lanteren, "The Potential for Entomophagous Parasites for Pest Control."

103. Paul R. Ehrlich and Peter H. Raven, "Butterflies and Plants: a Study in Coevolution," *Evolution* 18 (1964): 586-608.

104. Alan R. Putnam, "Allelopathic Chemicals: Nature's Herbicides in Action," *Chemical and Engineering News* 61 (1983): 34-45.

105. Marcos Kogan, "Plant Resistance in Pest Management," in Metcalf and Luckmann, *Introduction to Insect Pest Management*, pp. 93–134.

106. Bottrell, *Integrated Pest Management*.

107. M.S. Swaminathan, "Rice," *Scientific American* (1984): 81–93.

108. Bottrell, *Integrated Pest Management*.

109. *Ibid*.

110. Adkisson et al., "Controlling Cotton's Insect Pests."

111. Perry L. Adkisson and V.A. Dyck, "Resistant Varieties in Pest Management Systems," in Fowden G. Maxwell and Peter R. Jennings, eds., *Breeding Plants Resistant to Insects* (New York: John Wiley and Sons, 1980), pp. 233–51.

112. National Academy of Sciences/National Research Council, *Genetic Vulnerability of Major Crops* (Washington, D.C.: National Academy Press, 1972).

113. James L. Brewbaker, "Breeding for Disease Resistance," in Kommedahl and Williams, *Challenging Problems in Plant Health*, pp. 441–49.

114. J.A. Browning and K.J. Frey, "Multiline Cultivars as a Means of Disease Control," *Annual Review of Phytopathology* 7 (1969): 355–82.

115. M.S. Wolfe, "Genetic Strategies and their Value in Disease Control," in Kommedahl and Williams, *Challenging Problems in Plant Health*, pp. 461–73.

116. *Ibid*.

117. Swaminathan, "Rice."

118. Brewbaker, "Breeding for Disease Resistance."

119. Thomas C. Edens, Cynthia Fridgen, and Susan L. Battenfield, eds., *Sustainable Agriculture and Integrated Farming Systems* (East Lansing, Mich.: Michigan State University, 1985).

120. USDA Study Team, *Report and Recommendations on Organic Farming;* Robert C. Oelhaf, *Organic Agriculture* (Montclair, N.J.: Allanheld, Osmun and Co., 1978).

121. Miguel A. Altieri, *Agroecology: The Scientific Basis of Alternative Agriculture* (Berkeley, Calif.: University of California Division of Biological Control, 1983);

Stephen R. Gliessman, "Economic and Ecological Factors in Designing and Managing Sustainable Agroecosystems," in Edens et al., *Sustainable Agriculture and Integrated Farming Systems*, pp. 56–63; Patricia C. Matteson, Miguel A. Altieri, and Wayne C. Gagne, "Modification of Small Farmer Practices for Better Pest Management," *Annual Review of Entomology* 29 (1984): 383–402.

122. Robert Rodale, "The Past and Future of Regenerative Agriculture," in Edens et al., *Sustainable Agriculture and Integrated Farming Systems*, pp. 312–17.

123. Richard Harwood, private communication.

124. USDA Study Team, *Report and Recommendations on Organic Farming*; Oelhaf, *Organic Agriculture*; Robert Klepper et al., "Economic Performance and Energy Intensiveness on Organic and Conventional Farms in the Corn Belt," *American Journal of Agricultural Economics* 59 (1977): 1–12; J. Patrick Madden, Marlene K. Halvonson, and Stephen J. Dahm, "Regenerative Farming Systems," unpublished manuscript, 1984.

125. USDA Study Team, *Report and Recommendations on Organic Farming*; Batra, "Biological Control in Agroecosystems."

126. USDA Study Team, *Report and Recommendations on Organic Farming*.

127. Gliessman, "Economic and Ecological Factors in Designing and Managing Sustainable Agroecosystems."

128. Andow, "The Extent of Monoculture and its Effects on Insect Pest Populations."

129. Thomas C. Edens and Herman E. Koenig, "Agroecosystem Management in a Resource-Limited World," *BioScience* 30 (1981): 697–701; Edens and Haynes, "Closed System Agriculture."

130. K.E.F. Watt, "The Systems Point of View in Pest Management," in R.L. Rabb and F.E. Guthrie, eds., *Concepts of Pest Management* (Raleigh, N.C.: North Carolina State University, 1970), pp. 71–79; William G. Ruesink, "Status of the Systems Approach to Pest Management," *Annual Review of Entomology* 21 (1976): 27–44; J. Kranz and B. Hau, "Systems Analysis in Epidemiology," *Annual Review of Phytopathology* 18 (1980): 67–83.

131. Office of Technology Assessment, *Pest Management Strategies*.

132. H.T. Reynolds et al., "Cotton Insect Pest Management," in Metcalf and Luckmann, *Introduction to Insect Pest Management*, pp. 375–441.

133. Adkisson et al., "Controlling Cotton's Insect Pests."

134. Boza Barducci, "Ecological Consequences of Pesticides Used for the Control of Cotton Insects in Cañete Valley, Peru."

135. L. Brader, "Integrated Pest Control in the Developing World," *Annual Review of Entomology* 24 (1979): 225–54.

136. *Ibid.*

137. Qu Geping, "Biological Control of Pests in China."

138. W.H. Luckmann, "Integrating the Cropping System for Corn Insect Pest Management," in Metcalf and Luckmann, *Introduction to Insect Pest Management*, pp. 499–519.

139. Brader, "Integrated Pest Control in the Developing World."

140. B.A. Croft, "Apple Pest Management," in Metcalf and Luckmann, *Introduction to Insect Pest Management*, pp. 465–97; Whalon and Croft, "Apple IPM Implementation in North America."

141. B.A. Croft, J.L. Howes, and S.M. Welch, "A Computer-based Extension Pest Management Delivery System," *Environmental Entomology* 5 (1976): 20–24; S.M. Welch, "Developments in Computer-based IPM Extension Delivery Systems," *Annual Review of Entomology* 29 (1984): 359–81.

142. Whalon and Croft, "Apple IPM Implementation in North America."

143. Huffaker, *New Technology of Pest Control.*

144. Brader, "Integrated Pest Control in the Developing World."

145. Baugher, "Exposure to Pesticides During Application."

146. Wasserstrom and Wiles, *Field Duty.*

147. George E. Allen and James E. Bath, "The Conceptual and Institutional Aspects of Integrated Pest Management," *BioScience* 30 (1981): 658–64.

148. Frederick H. Buttel, "Biotechnology and Agricultural Research Policy: Emergent Issues"

79

(Ithaca, N.Y.: Cornell University Department of Rural Sociology, 1984), Bulletin No. 140.

149. Frederick H. Buttel and I. Garth Youngberg, "Implications of Biotechnology for the Development of Sustainable Agricultural Systems," in William Lockeretz, ed., *Environmentally Sound Agriculture* (New York: Praeger, 1983), pp. 377–400.

150. J. Artie Browning, "Goal for Plant Health in the Age of Plants: A National Plant Health System," in Kommedahl and Williams, *Challenging Problems in Plant Health*, pp. 45–57.

151. Manring, "Ecological Effects of Pesticides."

152. Mark G. Bookbinder, Douglas G. Baugher, and Karen R. Blundell, "Integrating IPM into the Pesticide Regulatory Process," consultant report to the U.S. Environmental Protection Agency from Dynamac Corporation, 1983.

153. Dover and Croft, *Getting Tough*.

154. Browning, "Goal for Plant Health."

155. Grace Goodell, "Challenges to International Pest Management Research and Extension in the Third World: Do We Really Want IPM to Work?" *Bulletin of the Entomological Society of America* 30 (1984): 18–26; Matteson et al., "Modification of Small Farmer Practices for Better Pest Management."

156. Ralph E. Hepp, *Alternative Delivery Systems for Farmers to Obtain Integrated Pest Management Services*, Agricultural Economics Report No. 298 (East Lansing, Mich.: Michigan State University, 1976).

157. Browning, "Goal for Plant Health."

158. Dover and Croft, *Getting Tough*.

159. Wasserstrom and Wiles, *Field Duty*.

160. Interagency IPM Coordinating Committee, *Report to the President: Progress Made by Federal Agencies in the Advancement of Integrated Pest Management* (Washington, D.C.: President's Council on Environmental Quality, 1980); Bottrell, *Integrated Pest Management*.

161. Dover and Croft, *Getting Tough*.

162. Wasserstrom and Wiles, *Field Duty*.

163. Daniel Benor and Michael Baxter, *Training and Visit Extension* (Washington, D.C.: The World Bank, 1984).

164. Robert Repetto, World Resources Institute report on pesticides subsidies, forthcoming.

# GLOBAL PESTICIDE USE Advisory Panel

*Lilia Albert*　Instituto Nacional de Investigaciones Sobre Recursos Bioticos, Mexico

*Randolph Barker*　Department of Agricultural Economics, Cornell University

*John C. Davies*　Department of Epidemiology & Public Health, School of Medicine, University of Miami

*Robert Oldford*　Union Carbide Corporation

*David Pimentel*　Department of Entomology, New York State College of Agriculture and Life Sciences, Cornell University

*Richard Sauer*　Institute of Agriculture, Forestry, and Home Economics, University of Minnesota

*Jacob Scherr*　Natural Resources Defense Council

*Alan Ternes*　Natural History Magazine

*Montague Yudelman*　World Resources Institute (formerly, Agriculture and Rural Development Office of the World Bank)

81

## World Resources Institute
1735 New York Avenue, N.W.
Washington, D.C. 20006 U.S.A.

*WRI's Board of Directors:*
Matthew Nimetz
*Chairman*
John E. Cantlon
*Vice Chairman*
John H. Adams
Robert O. Blake
John E. Bryson
Adrian W. DeWind
Marc J. Dourojeanni
John Firor
Curtis Hessler
Martin Holdgate
James A. Joseph
Alan R. McFarland
Robert S. McNamara
George P. Mitchell
Paulo Nogueira-Neto
Ruth Patrick
Amulya K.N. Reddy
Soedjatmoko
James Gustave Speth
M.S. Swaminathan
Mostafa K. Tolba
Russell E. Train
Arthur C. Upton
Melissa Wells
George M. Woodwell

James Gustave Speth
*President*
Jessica T. Mathews
*Vice President and Research
Director*
Kathleen Courrier
*Publications Director*

The World Resources Institute is a policy research center created in late 1982 with a grant from the John D. and Catherine T. MacArthur Foundation to help governments, international organizations, the private sector, and others address a fundamental question: How can we meet basic human needs and nurture economic growth without undermining the natural resources and environmental integrity on which life, economic vitality, and international security depend?

Independent and nonpartisan, World Resources approaches the development and analysis of resource policy options objectively, with a firm grounding in the natural and social sciences. Its research is aimed at providing accurate information about global resources and population, identifying emerging issues, and developing politically and economically workable proposals.

World Resources' research programs, based on global perspective and policy relevance, include living resources; agricultural resources; energy, climate, and industrial resources; pollution and health; and information, institutions, and governance. The Institute's research is carried out in part by its resident staff and in part by collaborators in universities and elsewhere.

World Resources is funded by private foundations, United Nations and governmental agencies, and corporations that share concern for resources, the environment, and sustainable economic development.

# WRI PUBLICATIONS ORDER FORM

| ORDER NO. | TITLE | QTY | TOTAL |
|---|---|---|---|
| S722 | *The Global Possible: Resources, Development, and the New Century* by the Global Possible Conference participants, $3.50* | | |
| S727 | *Improving Environmental Cooperation: The Roles of Multinational Corporations and Developing Countries* by WRI's Business Advisory Panel, $3.50* | | |
| S724 | *Getting Tough: Public Policy and the Management of Pesticide Resistance* by Michael Dover and Brian Croft, $3.50* | | |
| S725 | *Down to Business: Multinational Corporations, the Environment, and Development* by Charles S. Pearson, $3.50* | | |
| S714 | *Field Duty: U.S. Farmworkers and Pesticide Safety* by Robert F. Wasserstrom and Richard Wiles, $3.50* | | |
| S716 | *A Better Mousetrap: Improving Pest Management for Agriculture* by Michael J. Dover, $5.00* | | |
| S717 | *The American West's Acid Rain Test* by Philip Roth, Charles Blanchard, John Harte, Harvey Michaels, and Mohamed El-Ashry, $4.50** | | |
| S726 | *Helping Developing Countries Help Themselves: Toward a Congressional Agenda for Improved Resource and Environmental Management in the Third World* (a WRI working paper) by Lee M. Talbot, $10.00 | | |
| S744 | *Recommendations for a U.S. Strategy to Conserve Biological Diversity in Developing Countries* (a WRI working paper) by WRI's Biological Diversity Working Group, $10.00 | | |
| S745 | *The Conservation of Biological Diversity: A Report on U.S. Government Activities in International Resources Conservation* (a WRI working paper) by F. William Burley, $10.00 | | |
| | SUBSCRIBER OPTION A | | |
| | SUBSCRIBER OPTION B | | |

Name          (last)          (first)

Place of Work

Street Address

City/State          Postal Code/Country

Please send check or money order (U.S. dollars only) to WRI Publications, P.O. Box 620, Holmes, PA 19043-0620, U.S.A.

For prices and subscription rates see next page.

# BECOME A WRI SUBSCRIBER

■ **Option A.** Receive all WRI Policy Studies, all WRI research reports, and occasional publications for calendar year 1985. If you join in March, for example, you will immediately receive publications issued in January and February. $30.00. ($40.00 outside of the United States.)

■ **Option B.** Receive all WRI Policy Studies, all WRI research reports, all paperback books, and a 50 percent discount on all clothbound books published in calendar year 1985. $50.00. ($60.00 outside of the United States.)

All subscribers will also receive WRI's annual *Journal* free of charge.

**\*Policy Studies**
10-50—deduct $1.00 per copy
51-100—deduct $1.50 per copy

**\*\*Research Reports**
10-50—deduct $1.00 per copy
51-100—deduct $1.50 per copy

# WRI PUBLICATIONS ORDER FORM

| ORDER NO. | TITLE | QTY | TOTAL |
|---|---|---|---|
| S722 | *The Global Possible: Resources, Development, and the New Century* by the Global Possible Conference participants, $3.50* | | |
| S727 | *Improving Environmental Cooperation: The Roles of Multinational Corporations and Developing Countries* by WRI's Business Advisory Panel, $3.50* | | |
| S724 | *Getting Tough: Public Policy and the Management of Pesticide Resistance* by Michael Dover and Brian Croft, $3.50* | | |
| S725 | *Down to Business: Multinational Corporations, the Environment, and Development* by Charles S. Pearson, $3.50* | | |
| S714 | *Field Duty: U.S. Farmworkers and Pesticide Safety* by Robert F. Wasserstrom and Richard Wiles, $3.50* | | |
| S716 | *A Better Mousetrap: Improving Pest Management for Agriculture* by Michael J. Dover, $5.00* | | |
| S717 | *The American West's Acid Rain Test* by Philip Roth, Charles Blanchard, John Harte, Harvey Michaels, and Mohamed El-Ashry, $4.50** | | |
| S726 | *Helping Developing Countries Help Themselves: Toward a Congressional Agenda for Improved Resource and Environmental Management in the Third World* (a WRI working paper) by Lee M. Talbot, $10.00 | | |
| S744 | *Recommendations for a U.S. Strategy to Conserve Biological Diversity in Developing Countries* (a WRI working paper) by WRI's Biological Diversity Working Group, $10.00 | | |
| S745 | *The Conservation of Biological Diversity: A Report on U.S. Government Activities in International Resources Conservation* (a WRI working paper) by F. William Burley, $10.00 | | |
| | SUBSCRIBER OPTION A | | |
| | SUBSCRIBER OPTION B | | |

Name          (last)                              (first)

Place of Work

Street Address

City/State                                      Postal Code/Country

Please send check or money order (U.S. dollars only) to WRI Publications, P.O. Box 620, Holmes, PA 19043-0620, U.S.A.

For prices and subscription rates see next page.

# BECOME A WRI SUBSCRIBER

■ **Option A.** Receive all WRI Policy Studies, all WRI research reports, and occasional publications for calendar year 1985. If you join in March, for example, you will immediately receive publications issued in January and February. $30.00. ($40.00 outside of the United States.)

■ **Option B.** Receive all WRI Policy Studies, all WRI research reports, all paperback books, and a 50 percent discount on all clothbound books published in calendar year 1985. $50.00. ($60.00 outside of the United States.)

All subscribers will also receive WRI's annual *Journal* free of charge.

**\*Policy Studies**
10-50—deduct $1.00 per copy
51-100—deduct $1.50 per copy

**\*\*Research Reports**
10-50—deduct $1.00 per copy
51-100—deduct $1.50 per copy

# WRI PUBLICATIONS ORDER FORM

| ORDER NO. | TITLE | QTY | TOTAL |
|---|---|---|---|
| S722 | *The Global Possible: Resources, Development, and the New Century* by the Global Possible Conference participants, $3.50* | | |
| S727 | *Improving Environmental Cooperation: The Roles of Multinational Corporations and Developing Countries* by WRI's Business Advisory Panel, $3.50* | | |
| S724 | *Getting Tough: Public Policy and the Management of Pesticide Resistance* by Michael Dover and Brian Croft, $3.50* | | |
| S725 | *Down to Business: Multinational Corporations, the Environment, and Development* by Charles S. Pearson, $3.50* | | |
| S714 | *Field Duty: U.S. Farmworkers and Pesticide Safety* by Robert F. Wasserstrom and Richard Wiles, $3.50* | | |
| S716 | *A Better Mousetrap: Improving Pest Management for Agriculture* by Michael J. Dover, $5.00* | | |
| S717 | *The American West's Acid Rain Test* by Philip Roth, Charles Blanchard, John Harte, Harvey Michaels, and Mohamed El-Ashry, $4.50** | | |
| S726 | *Helping Developing Countries Help Themselves: Toward a Congressional Agenda for Improved Resource and Environmental Management in the Third World* (a WRI working paper) by Lee M. Talbot, $10.00 | | |
| S744 | *Recommendations for a U.S. Strategy to Conserve Biological Diversity in Developing Countries* (a WRI working paper) by WRI's Biological Diversity Working Group, $10.00 | | |
| S745 | *The Conservation of Biological Diversity: A Report on U.S. Government Activities in International Resources Conservation* (a WRI working paper) by F. William Burley, $10.00 | | |
| | SUBSCRIBER OPTION A | | |
| | SUBSCRIBER OPTION B | | |

Name                    (last)                                        (first)

Place of Work

Street Address

City/State                                                    Postal Code/Country

Please send check or money order (U.S. dollars only) to WRI Publications, P.O. Box 620, Holmes, PA 19043-0620, U.S.A.

For prices and subscription rates see next page.

# BECOME A WRI SUBSCRIBER

■ **Option A.** Receive all WRI Policy Studies, all WRI research reports, and occasional publications for calendar year 1985. If you join in March, for example, you will immediately receive publications issued in January and February. $30.00. ($40.00 outside of the United States.)

■ **Option B.** Receive all WRI Policy Studies, all WRI research reports, all paperback books, and a 50 percent discount on all clothbound books published in calendar year 1985. $50.00. ($60.00 outside of the United States.)

All subscribers will also receive WRI's annual *Journal* free of charge.

**\*Policy Studies**
10-50—deduct $1.00 per copy
51-100—deduct $1.50 per copy

**\*\*Research Reports**
10-50—deduct $1.00 per copy
51-100—deduct $1.50 per copy